The Volunteer Fire Company

ERNEST EARNEST

THE VOLUNTEER FIRE COMPANY

Past and Present

𝕾𝔇 STEIN AND DAY/Publishers/New York

First published in 1979
Copyright © 1979 by Ernest Earnest
All rights reserved
Designed by L.E. Heblin
Printed in the United States of America
Stein and Day/*Publishers*/Scarborough House,
Briarcliff Manor, N.Y. 10510

Library of Congress Cataloging in Publication Data

Earnest, Ernest Penney, 1901–
    The volunteer fire company book.

    1. Fire-departments—United States.   I.   Title.
TH9503.E37   363.3′7′0973   78-8785
    ISBN 0-8128-2532-2

Dedicated to the
Gladwyne Fire Company

ACKNOWLEDGMENTS

This book was prepared with a grant-in-aid from the American Philosophical Society.
I also gratefully acknowledge the information supplied by the following persons and organizations.

LIBRARIES AND MUSEUMS
The Haverford College Library
The Hershey Museum
The Home Insurance Company
The Insurance Company of North America Archives
The Library Company of Philadelphia
The Mutual Assurance Company
The Pennsylvania Historical Society
Temple University: Paley Library
The William Penn Memorial Museum, Harrisburg, Pa.

ASSOCIATIONS, CORPORATIONS, AND STATE SERVICES
American Insurance Association
American La France
Arkansas Training Academy
California State Fire Marshal
Florida, State Fire Marshal
Friendship Veteran Fire Engine Association, Alexandria, Va.

Insurance Information Institute
Iowa State University: Fire Service Extension
Lower Merion Township: (Pa.) Department of Fire
Michigan State Police Fire Marshal
Minnesota Department of Public Safety
National Fire Protection Association
Nebraska Fire Training Academy
Nebraska State Volunteer Firemen's Association
North Dakota State Fire Marshal
Oregon State Fire Marshal
Pennsylvania State University: Pennsylvania Technical Assistance
Program
The *Philadelphia Bulletin*
Snorkel, A-T-O, Inc.
Virginia State Department of Education
Volunteer Firemen's Insurance Services, Inc.
Waterous Company

CHIEFS AND FIRE COMPANIES
Boniface Aiu, Honolulu, Hawaii
Milton G. Alward, Sr., Gladwin, Mich.
Howard A. Blythe (member) Old Greenwich, Conn.
R.F.C. Brewer, Maine
Tom Chapman, Wayne, N.J.
Steve Childers, Fairbanks, Alaska
Charles E. Chronister, York Springs, Pa.
J. Collins, Franklin Park, Ill.
Robert K. Condit, Sewickley, Pa.
Cooperstown, N.Y. Fire Department
Elmer L. Cotner, Washingtonville, Pa.
Richard E. Dean, Avalon, N.J.
Frank DeBloise, Guyasuta Fire Dept., Pittsburgh, Pa.
Thomas Derrick, Lake Geneva, Wis.
Dennis A. Dewalt, Brentwood, Pa.
Reginald Donette, Lewistown, Maine
Homer D. Elicker, Harrisburg, Pa.
David L. Frey, Williamsport, Pa.
Jack R. Gagne, Deerfield-Bannockburn, Ill.
Francis X. Gersenhoffend, Waverly, Tenn.

Corville Giner, Tuckahoe, N.J.
R. Gorland, Gladwater, Tex.
Jack M. Hamilton, Prescott, Ariz.
Leo Heine, Chilton, Wis.
Willard Herring, Goldsboro, N.C.
R. Holman, Jr., Augusta, Maine
Jeneau Fire Department, Alaska
George Johnson, Jamestown, N.Dak.
Mark D. Johnson, Honolulu, Hawaii
M. O. Johnson, Point Pleasant, W.Va.
Robert L. Kins, Wellsburg, W.Va.
La Crosse, Wis. Fire Department
S. J. La Scola, Takoma Park, Md.
G. A. Lemon, New Kensington, Pa.
Robert Little, York, Pa.
Tommy W. Livingston, Hartsville, S.C.
J. Lloyd, Martinsville, Va.
Elsie L. Lowe (editor) Townsend, Mass.
G. Moser, Hickory, N.C.
Roger C. O'Donnell, Covington, Ohio
Robert D. Porter, Forest Grove, Pa.
R. L. Raynes, Eleanor, W.Va.
W. Donley Reed, Monroeville, Pa.
Riverhead, N.Y. Fire Department
Salvadore Romero, Clifton, Ariz.
W. Rosenfelder, Geneva, Ill.
James Scriver, Oscoda, Mich.
Ignas Servia, Tarrytown, N.Y.
Sidney, Mont. Fire Department
John Sokol (editor) Summit Fire Department, N. Versailles, Pa.
John Stacey, Bellevue, Md.
Calvin Studebaker, Fredonia, Kans.
Thomas Styche, Beuna Vista, Pa.
Charles C. Talliat, Centerville, Iowa
L. B. Topliff, Plattsburg, Nebr.
Maurice Van Hamme, Belle Plaine, Iowa
Richard Wagner, Hamilton Township, Allison Park, Pa.
Robert N. Waugh, Orange, Va.
N. R. Wiles (secretary) Nebraska City, Neb.

X   ACKNOWLEDGMENTS

Rey Gene Williams, Bowie, Tex.
Walter Winston, Ketchikan, Alaska
Edward J. York, Radnor Fire Co., Wayne, Pa.
W. F. Zaiber, Burlington, Iowa

Alfred A. Knopf, Inc. for permission to use the lines from "A Postcard from the Volcano," of *The Collected Poems of Wallace Stevens.*

Lines from "Speech to Those Who Say Comrad," from *New and Collected Poems 1917-1976* by Archibald MacLeish, copyright 1976. Reprinted by permission of Houghton Mifflin Company.

CONTENTS

List of Illustrations

The Volunteer Fire Company

I

AN ENDURING INSTITUTION

THE VOLUNTEER FIRE COMPANY is a characteristically American institution. At various times there have been volunteer firemen in other countries, but nowhere outside of the United States have volunteer fire companies been so ubiquitous or played such a prominent part in social history. The American volunteer company, unlike those elsewhere, was almost always organized by concerned citizens rather than by a governmental body, and to a large extent the American companies have been self-governing, establishing their own rules and electing their own officers.

Volunteer fire companies appeared early on the American scene and continue to flourish, not only in small communities but in some cities. Reading, Pennsylvania, a city of 88,000, is the largest municipality protected by a volunteer

1

fire department, but Bloomington, Minnesota, with a population of 80,000, is not far behind. The spread of dates is
significant. Reading's Rainbow Company, probably the oldest volunteer unit in continuous service, dates from 1773;
the Bloomington Fire Department was founded in 1947.

In the United States there are over 24,000 volunteer and
part-paid fire departments.* For some reason Pennsylvania
has the largest number of such departments—2,980, or 99.7
percent of all departments in the state. It must be recognized
that large cities like Philadelphia and Pittsburgh have departments made up of many separate companies, whereas a
small town department will have only one or two companies, Other states with a large number of volunteer or
part-paid departments are Minnesota with 751, or 97.8 percent of all fire companies; New York with 1,850 or 96.5
percent; Ohio with 1,178 or 93.6 percent; and Illinois with
1,029 or 85 percent.** On the other hand, all departments
in Hawaii and the District of Columbia are paid.

The number of fire fighters in the United States in 1972 as
estimated by the National Fire Protection Association was
about 2 million, of whom 250,000 were full-time paid men,
the rest volunteers.*** In 1958, when latest figures were

---

*The term part-paid means those departments with a nucleus of paid fire fighters
and a much larger number of volunteer or on-call fire fighters. Departments with a
large number of paid fire fighters with some volunteers assisting are properly called
"predominently paid." From very early times, as will appear, volunteers sometimes
received pay or other recompense for fire duty. Today there is a wide variety of
practice from a small fixed payment or hourly rate for fire calls to an annual
stipend of perhaps $500. Most volunteers receive no pay at all.

As used here *volunteer* will mean a person whose primary occupation is not fire
fighting, but who is a member of a fire company and who leaves home or occupation to answer alarms.

**As the National Fire Protection Association stated in 1972, "Accurate figures on
fire departments and firefighters do not exist." Thus the IAFC *Newsletter* gives the
figure 1,728 for volunteer departments in New York; the Volunteer Firemen's
Insurance Service gives 1,899; a letter of 1976 from the Department of State gives
1,850. Wherever possible I have used the latest figures available.

***As of 1972 and earlier the *fireman* was applicable; more recently women have
begun to enter the fire service, so the later term must be *fire fighter*.

available, Pennsylvania companies had 432,000 members, of whom 102,000 were active fire fighters. As of 1976 Virginia had an estimated 20,000 volunteers and Minnesota 15,000.

And far from fading away, new volunteer fire companies are being formed in recently settled communities, and counties are encouraging the formation of regional fire departments to cover rural areas.

As long ago as the 1830s Alexis de Toqueville recognized the American habit of forming associations for all sorts of purposes:

Americans of all ages, all conditions, and all dispositions constantly form associations. They have not only commercial and manufacturing companies, in which all take part, but associations of a thousand kinds, religious, moral, serious, futile, general or restricted, enormous or diminutive. The Americans make associations to give entertainments, to found seminaries, to build inns, to construct churches, to diffuse books, to send missionaries to the antipodes; in this manner they found hospitals, prisons, and schools. If it is proposed to inculcate some truth or to foster some feeling by encouragement of a great example, they form a society. Whenever at the head of some new undertaking you see the government in France, or a man of rank in England, in the United States you will be sure to find an association.*

That this national trait exists a century and a half later is obvious in such organizations as Rotary, Kiwanis, and Lions Clubs; the PTA and the ubiquitous suburban civic associations, garden clubs, and the more ephemeral groups formed

---

*As will appear, Napoleon put the Paris fire department under government control, and in England estate brigades were set up by the lord of the manor.

to save a historic building, create a park or a playground. The list is endless.

However, in many ways the volunteer fire company is unique in that it performs a service that in most other areas of life has been delegated to paid professionals. One needs only to mention areas that have become professionalized: no longer, as a rule, do neighborhood women nurse the sick, act as midwives, or lay out the dead. It is now newsworthy when a group of unpaid men build a house or barn for a burned-out neighbor. Householders are no longer expected to clean the street in front of their dwellings.

The school has long since taken over training the skills a boy or girl will need as an adult, and increasingly child care is delegated to the nursery school and the kindergarten. Only in rural areas do people dispose of their own garbage, and even farm families buy their bread and meat from the assembly line. All this is so obvious that it is taken for granted. But men, and recently women also, still turn out of warm beds in January at the sound of the bull whistle or the siren, and dash out to devote unpaid hours to save lives and property.

It is easy to understand the original impulse for the formation of volunteer fire companies: they were part of the do-it-yourself tradition that developed in a new country lacking in long-established lines of authority. In many ways they were concomitant with the growth of democratic institutions such as the town meeting, the local militia company, the colonial assembly, and of course the Continental Congress.

By contrast, the first fire companies in London and other British cities were the creations of insurance companies and confined their efforts to saving property designated by a company fire mark. In the first two decades of the nineteenth century the first volunteer fire brigades were formed in England, usually under the aegis of the municipality or the local gentry. Big estates maintained their own brigades made up of footmen, gamekeepers, stablemen and house servants. It was not until 1814 that Ashford in Kent formed one of the first volunteer brigades without official backing or payment, and they provided themselves with uniforms and equipment at their own expense or through subscriptions from gentry and trades people.

Although volunteer brigades continued in England for 120 years, their heyday was not reached until the mid-nineteenth century when insurance brigades began to disappear.

In Paris a civilian fire service was formed in 1716 but went out of existence following a fire and panic on the night of July 1, 1810 during a dance at the Austrian embassy in honor of Napoleon's marriage. He placed the responsibility for the panic and deaths of ten persons upon the civilian force, and the following year created the military Battalion of Fire Fighters of the City of Paris. The department is now a regiment of the French Army Engineers.

In 1715 a Dutch merchant was shocked at seeing 50 houses burnt at Wapping. The London fire fighting was primitive compared to Amsterdam's, where a city brigade with 60 engines had existed for over 50 years. In Germany firemen are organized on severely official lines.

For the volunteer fire company there were—and still are—unique economic reasons. Except in large cities fires were and are of sporadic occurrence. When one happens, a sizable body of men may be needed. This was especially true in the days of leather buckets and hand-pumped engines. Thus, in many places a small number of paid firemen would be idle much of the time and would be inadequate to handle a large conflagration. This will be discussed later in more detail.

The psychological reasons for the fascination of a fire are obscure. It is easy enough to account for the atavistic love for a fire on the hearth, but less easy to explain the lure of a big blaze. There may be something of the arsonist in most people.

Thus Thoreau wrote

> If I should only give a few pulls at the parish bellrope, as for a fire . . . there is hardly a man on his farm in the outskirts of Concord, notwithstanding that press of engagements which was his excuse so many times this morning, nor a boy, nor a woman, I might almost say, but would forsake all and follow that sound, not mainly to save property from the flames, but, if we will confess the truth, much more to see it burn, since burn it must, and we, be it known, did not set it on fire—or to see it put out, and have a hand in it, if that is done as handsomely; yes, even if it were the parish church itself.

There, by the way, is a description of the two types at every fire: the fire fighters and the onlookers who clutter up the scene and get in the way. Volunteer companies can handle a big blaze, but it often requires professional police to handle the rubbernecks.

Of course, any kind of disaster will attract a crowd, but a big fire is an especially dramatic and ongoing spectacle. The huge popularity of the Currier and Ives sequence *The Life of a Fireman* is testimony to the public's fascination with fires and fire fighting past and present. Toy fire engines have been children's favorites for over 100 years and kids love to hang around a firehouse and, if permitted, clamber on the engines. When a firehouse is used as a polling place, the voters are likely to inspect the apparatus and ask questions about it.

In cities the firebuff is a ubiquitous type, but he existed long before he was given a quasi-official status. The wealthy

New York lawyer and diarist, George Templeton Strong (1820-1875), recorded running to fires from the time of his student days at Columbia University until his death at 55. In his younger days his father also got out of bed, even in January, to follow the engines.

At a fire in the 1840s where two men were killed, Strong noted: ". . . the flames flashed up beautifully through the smoke." Seventeen years later he described the burning of Harper and Brothers as "more than commonly dramatic and splendid . . . . Walls thundering down at intervals." And in 1860 at a tenement fire where 30 died: "Steam fire engines are a new element in our conflagrations and an effective one contributing to the *tout ensemble* a column of smoke and sparks and a low shuddering throbbing bass note, more impressive than the clank of the old-fashioned machines. . . ." It would seem that Strong was more of a connoisseur than a humanitarian.

In recent years the most noted firebuff was Mayor Fiorello LaGuardia, who certainly never lost votes by appearing at fires. Philadelphia's Mayor Frank Rizzo got a lot of publicity in 1976 by breaking a leg while getting away from a refinery fire he had got too close to.

Undoubtedly, the drama of fire fighting is one reason why men and women join volunteer fire companies, but this is only part of the story. There is nothing heroic about hours of drill or dramatic about packing wet and often dirty hose for an hour or two. The life of a fire fighter has more drudgery than drama.

II

IN THE BEGINNING

BENJAMIN FRANKLIN DID NOT found the first American fire company. This is an enduring legend especially cherished in Philadelphia. However, the distinction of being first belongs to Boston, where an engine was imported from England in 1679, twenty years before Paris had one. The selectmen seem to have set up some sort of fire company, because seven years later they found it necessary to revive it and place it under the command of Henry Deering. The company met on the last Monday of each month to "Exercise themselves in the use of Sd. Engine."

Following a fire in 1702 which destroyed eight warehouses, the old engine was repaired and a new one ordered from England. In 1707 another engine was acquired, was well-housed, and a second company of 20 men was formed.

8

After another big fire in 1711 which destroyed 80 buildings, the city bought three more engines.

Despite all this, the handling of fires was far from satisfactory. Months after the 1711 fire the justices were still advertising for the return of property taken from homes during "the late fire." In Boston, as elsewhere, looting at fires was a constant problem. A solution was suggested in a communication to the *News Letter* for February 6, 1715/16 in which the writer blamed the lack of success in extinguishing fires and the general pilfering which accompanied them on the "great Crowd and Confusion that often falls out at a Fire; where no Body Governs, no Body will Obey, very few will Work, and a great many Lookers on . . . only incumber the Ground." The writer proposed the adoption of a system like that in Amsterdam, Holland.

This seems to have been the inspiration for the formation of the first volunteer fire company in America, the Boston Fire Society of 1717, 19 years before Franklin's Union Fire Company in Philadelphia. The Fire Society was composed of 20 men, each of whom agreed to bring two buckets and two large bags to every fire. The bags were for the purpose of salvage: goods were taken to a designated spot where someone stood on guard. The rules of the Society furnished the model for the Union Fire Company and those which followed it.

The second American city to acquire a fire engine was Charles Town, South Carolina in 1713. This city had experienced several disastrous fires, one in 1698 destroying much of the town; another a year later causing a loss estimated at §30,000; and one in 1700 reducing most of the town to ashes.

Fire engines are, of course, not the whole story of early fire protection. In 1647, a year after he landed, Peter Stuyvesant appointed four fire wardens to inspect chimneys of the thatched-roofed houses and to levy a fine of three guilders for each unswept chimney. The money was used to import leather buckets, hooks, and ladders. In Boston, after a disastrous fire in 1754 which killed three people, a set of laws was passed providing:

1. Each house was to have a ladder or ladders long enough to reach the ridge pole.
2. Every house to have a 12-foot-pole with a swab. [For chimney fires, a constant hazard.]
3. Six good and long ladders were to be hung outside the meeting house.
4. Four good, strong iron crooks with chains and ropes to hang outside the meeting house.
5. The town was to establish a cistern or conduit at the corner of State and Exchange Streets.
6. It was forbidden to build a fire within 3 rods of any building.
7. For arson the penalty was death.
8. It was illegal to carry burning brands from one house to another, except in a safe vessel.
9. The town was to establish a night patrol.
10. No fires were permitted in vessels at the wharf.

Other cities passed similar ordinances. In 1698 and again in 1704 the South Carolina assembly required that all chimneys must be constructed of brick or stone and all wooden chimneys were to be demolished within a month. In New York, 1691, Dick Vandenburg, a bricklayer, headed a committee to inspect fireplaces and chimneys to see that they were kept clean. In 1697 the alderman of each ward and an assistant were made responsible for hearths and chimneys therein. The Philadelphia Corporation, setting a precedent to be followed for two centuries, failed to enforce the fire laws with the result that the Provincial Assembly levied a fine of 40 shillings for uncleaned chimneys.

Then as now highly combustible materials were a menace. In the eighteenth century the hazards included pitch, paint, gunpowder, and the shavings in woodworking shops; in the nineteenth there were the added hazards of stores of turpentine and coal oil; and now there is a whole arsenal of explosive, poisonous, and flammable substances, both stored and in transit. A law of 1701 in Philadelphia prohibited the possession of more than 6 pounds of gunpowder unless stored 40 perches from any building.

However, in the days of open fires, candles, lamps, and lanterns the numerous laws for fire protection did not prevent frequent fires, many of which because of inadequate means of control spread over a large area. In Boston alone there were large conflagrations in 1654, 1676 (46 houses); 1679 (80 houses); 1711 (100 buildings); and 1760 (400 buildings). As has been mentioned, Charleston, South Carolina experienced several disastrous fires in the late seventeenth and early eighteenth centuries.

The American custom of building with wood was, of course, a major reason for the prevalence and spread of fires in this country. By contrast, in London after the great fire of 1666 the law required masonry construction. Mediterranean buildings of stone with tiled roofs are obviously not very flammable.

Before and after the introduction of fire engines citizens were often required to keep leather buckets ready. In 1686 New York ordered that "every person having two chimneys to his house provide one bucket." A law of 1696 in Philadelphia provided that householders must have leather buckets and ladders. Justices of the peace were to provide six or eight good hooks to be used in pulling down houses, and were given authority to blow up buildings if necessary.

The requirement that householders have buckets and ladders suggests that early fire fighting was less an organized activity than a kind of free-for-all. Even after the introduction of fire engines operated by organized companies, citizens were pressed into service to form two chains, one to pass full buckets from a well or creek, the other to return the empties. Persons not actively engaged in the work were required to answer the demand: "Throw out your buckets." Thus in 1729 James Barrett was paid £6 for twelve buckets taken from him at the last fire on Chestnut Street" in Philadelphia. The names of the owners were always painted on the buckets. After a fire the watchman would call out. "Hear ye, oh, I pray ye, claim your buckets."

It is clear that fire fighting of a rather haphazard nature antedated the formation of organized fire companies. As has been noted, the first company in Boston had died out and had to be revived in 1686. Philadelphia got its first fire engine in 1719, seventeen years before Franklin organized the Union Fire Company. As might be expected when everybody's business is nobody's business, the apparatus was neglected. By 1726 the engine was "much out of repair" and a committee of aldermen was instructed to view it and "think of a proper place to preserve it from the weather."

In the absence of an organized company, the Council appointed an engineer to be in charge of the machine. Realizing the inadequacy of fire protection for the rapidly growing city, the Council in 1730 agreed to purchase three more engines, one at about £50, one at £35, and one at about £20, all from England, along with 200 leather buckets. An additional 200 plus 20 ladders were purchased locally. To cover the cost a special tax was levied.

Apparently all this equipment was to be used by any citizens who happened to be available. It will be remembered that before the formation of a fire company Boston had provided that ladders, hooks, and chains be hung outside the meeting house. The pumps were, of course, operated by men, but women and children often joined the bucket line to keep the engine full.

This catch-as-catch-can method of fighting fires often continued long after the formation of regular companies. In Chicago, until the advent of a paid department in 1858, all citizens could be called upon to fight fires in the event of special need. This enlistment of ordinary citizens explains the misadventure of young Henry James at a fire in Newport in 1860. When the young men standing nearby were asked to lend a hand, Henry could not avoid joining in:

> . . . the willing youths, all round were mostly starting to their feet, and to have trumped up a lameness at such a juncture could be made to pass in no light for graceful. Jammed into the acute angle between two high fences, where the rhythmic play of my arms in tune with several other pairs, but at a dire disadvantage of position, induced a rusty, a quasi-extemporized old engine to work and a saving stream to flow, I had done myself, in the face of a shabby conflagration, a horrid even if obscure hurt. . . .

Before the days of motorized apparatus a spare pump or hose carriage might be kept in some barn or shed at a distance from a regular firehouse. As late as 1920 in one of those elongated Pennsylvania small towns, a two-wheeled hose cart was available to any men who happened to be in the neighborhood, which was a mile from the main firehouse. And, of course, in the era of leather buckets these were available in almost every household, and in some places were required by law. Just as today any citizen with a garden hose will usually squirt it on a nearby fire, so anyone with a bucket was almost sure to use it—probably with equal ineffectiveness.

Obviously, as towns grew into cities with the consequent crowding of buildings, more organized efforts were needed. Boston, with its first fire company in 1679, was far ahead of other cities in this regard. The reorganized company of seven years later adopted that most important feature of effective firefighting—regular drill to "Exercise themselves in the use of Sd. Engine." As will appear, this has become ever more important with the increasing complexity of fire apparatus. The point is that from the very beginning volunteer companies adopted this practice and enforced participation by their members. Then as now the untrained citizen acting on his own is at best a nuisance at a fire, and at worst a menace to himself and others.

In 1736 Newport, Rhode Island adopted the requirement that engine companies inspect and try their apparatus regularly. By that time the town had three engines. Ten years before a group of prominent citizens had founded a Fire Club similar to the Boston Fire Society, and had bought buckets and an engine. After a serious fire in 1730 Newport's richest citizen, Col. Godfrey Malbone presented a new suction-type engine made by Newsham and Ragg in London.*

At first the town took over the care of the engine, but neglected it. Three years after its purchase the Fire Club complained that the engine and buckets "are not as carefully looked after as they ought to be," whereupon the town council asked that the club managers appoint persons to look

---

*A suction engine could draw water directly from a pond, cistern, or well, thus making a bucket line unnecessary.

after the equipment "under such Regulations as they Shall think Proper."

Obviously both Boston and Newport had fire companies before Philadelphia. In that city a letter in the *Pennsylvania Gazette* (Franklin's paper) February 4, seventeen thirty-four/thirty-five stated:

As to our Conduct in the Affair of Extinguishing Fires, tho' we do not want Hands or Good-will, yet we seem to want Order and Method. . . .

The writer went on to propose:

. . . for our Imitation, the Example of a City in a neighboring Province. There is as I am well informed, a Club or Society of active Men belonging to each Fire Engine, whose Business is to attend all Fires with it whenever they happen; and to work it once a Quarter, and to see it kept in order. Some are to handle the Firehooks, and other the Axes, which are always kept with the Engine; and for this service they are consider'd in Abatement or Exemption in the Taxes. In Time of Fire, they are commanded by Officers appointed by Law, called *Firewards,* who are distinguished by a Red Staff of five Feet long, headed by a Brass Flame of 6 Inches; and being Men of Prudence and Authority, they direct the opening and stripping of Roofs by the Ax-Men, the pulling down burning Timbers by the Hookmen, and the playing of the Engines. . . .

This description of regular meetings, drills with the apparatus, the divisions of function, and chain of command shows that the Fire Society of Boston had set the pattern for fire companies ever since.

(1)

The seventh Day of December, in the Year of our Lord One thousand seven hundred and thirty Six, We whose Names are hereunto subscribed, reposing special Confidence in each others Friendship, Do, for the better preserving our Goods and Effects from Fire, mutually agree in manner following, that is to say.

I. 1. That we will each of us at his own proper Charge provide Two Leathern Buckets, and Four Baggs of good Ozenabrigs or wider Linnen, Whereof each Bagg shall contain four Yards at least, And shall have a running Cord near the Mouth; Which said Buckets and Baggs shall be marked with the Initial Letters of our respective Names & Compª. Thus [A.B. & Compª.] and shall be apply'd to no other Use than for preserving our Goods & Effects in Case of Fire as aforesaid.

II. 2. That if any one of us shall fail to provide and keep his Buckets and Baggs as aforesᵈ he shall forfeit and pay unto the Clerk for the Time being, for the Use of the Company, The Sum of Five Shillings for every Bucket and Bagg wanting.

3. That if any of the Buckets or Baggs aforesᵈ shall be lost or damaged at any Fire aforesᵈ The same shall be supplyed and repaired at the Charge of the whole Company.

4. That we will all of us, upon hearing of Fire breaking out at or near any of our Dwelling Houses, immediately repair to the same with all our Buckets and Baggs, and there employ our best Endeavours to preserve the Goods and Effects of such of us as shall be in Danger by Packing the same into our Baggs: And if more than one of us shall be in Danger at the same time, we will divide our selves as near as may be to be equally helpful. And to prevent suspicious Persons from coming into, or carrying any Goods out of, any such House, Two of our Number shall constantly attend at the Doors until all the Goods and Effects that can be saved shall be secured in our Baggs, and carryed to some safe Place, to be appointed by such of our Company as shall be present, Where one or more of us shall attend them till they can be conveniently delivered to, or secured for, the Owner.

5. That we will meet together in the Ev'ning of the last second Day of the Week commonly called Monday, in every Month, at some convenient Place to be appointed at each Meeting, To consider of what may be further useful in the Premises; And whatsoever shall be expended at every meeting aforesᵈ shall be paid by the Members met. And if any Member shall neglect to meet as aforesᵈ, he shall forfeit and pay the Sum of One Shilling.

6. That we will each of us in our turns, according to the Order of our Subscriptions, serve the Company as Clerk for the Space of one Month, viz. That Member whose Name is hereunto first subscribed shall serve first, and so on to the Last. Whose Business shall be to inspect the Condition of each of our Buckets and Baggs, and to make Report thereof at every monthly Meeting aforesᵈ.

Above and opposite top: Articles of the Union Fire Company with Benjamin Franklin's signature, dated December 7, 1736 (*Courtesy Pennsylvania Historical Society*)

Hibernia Fire Engine Company No. 1 (*Courtesy INA Corporation Museum*)

The uniform of Hibernia Engine Company No. 1 (*Courtesy INA Corporation Museum*)

Above: 1764 pumper made in England, purchased secondhand in 1812 by the Manheim, Pennsylvania, Fire Company *(Courtesy Hershey Museum of American Life)*

Napoleon Earnest

Left: Friendship Veterans Fire Engine Company, organized by George Washington in 1774. General Washington was an active member.

Below: Citizen engine dated 1788 from Hummelstown, Pennsylvania. Now in the William Penn Memorial Museum, Harrisburg, Pennsylvania. The writer's grandfather Napoleon Earnest, a Hummelstown fireman in the 1870s, remembered pumping it. *(Courtesy William Penn Memorial Museum)*

That the letter, although signed A. A., was almost certainly written by Franklin is clear from the *Autobiography* in which he states, "I wrote a paper (first read in Junto, but it was afterward published) on the different accidents and carelessness by which houses were set on fire, with cautions against them. This was much spoken of as a useful piece, and gave rise to a project, which soon followed it, of forming a company for the more ready extinguishing of fires, and mutual assistance in removing and securing goods when in danger." This exactly describes the letter in the *Pennsylvania Gazette* in which the proposal for a fire company was prefaced by a discussion of fire prevention, and included remarks on salvage.

Although the Boston Articles of Agreement have been lost, it is probable that those of the Union Fire Company were similar. They provided that each member furnish six leather buckets and two linen bags for removing goods from a burning building. All these had to be maintained in good condition. The members were to meet once a month for a social evening and discussion of fire fighting methods and problems. The fines levied for absences were to be used to buy additional equipment.

As the membership was limited to 30, the applicants soon became so numerous that some were advised to form a second company. This was the Fellowship Fire Company organized by Franklin's rival, Andrew Bradford, in 1738. This was followed by the Hand in Hand 1742, and by another in 1743. By the time of the Revolution there were 22 fire companies in Philadelphia and Germantown. Writing in 1788, Franklin questioned if any city in the world was better provided with the means to stop fires from becoming conflagrations. He claimed that since the institution of fire companies the city had never lost more than one or two houses at a time.

Franklin's comparison of Philadelphia with other cities was not an idle boast. For eighteenth century Britain, Walford's *Insurance Cyclopaedia* listed well over 300 great conflagrations. In America during that century a wind-blown fire in New York in 1741 consumed the governor's house, the barracks, and the secretary's office. Another, possibly of incendiary origin, in 1776 destroyed 493 houses—about one-fourth of the city. A fire in Charleston, 1731, gutted most of the town; another in 1795 destroyed buildings on 253 acres; one in 1798 in Wilmington, North Carolina destroyed all but 12 houses.

New York had been slow in providing for the extinguishing of fires. In 1729 a Dutchman writing in the *New York Gazette* complained of the lack of foresight by the English. Many people rushed to fight a fire but they were entirely without equipment. After praising the excellent apparatus in London he went on: "Nay, the city of Philadelphia (as young as it is) has had two Fire Engines for several years past; and it is a wonder to many that this city should so long neglect the getting of one or more of them." A year later the Corporation finally got around to levying a property tax, which among other things, financed the purchase of two Newsham engines * with suctions, plus leather hose, caps, and other material.

These arrived in 1731, but the public was not satisfied. One writer said that more engines were needed, and that there should be regular drills. This led the corporation to action: they purchased a speaking trumpet for use at fires. Then in 1737 the Assembly authorized the Corporation to appoint 30 strong men to "have the Care, management,

---

*Until the invention of the Newsham engine in 1725, the early fire engines were less efficient than the Roman ones designed by Clesibius of Alexandria in the third century B.C. Fragments found in England show that they had two bronze cylinders, pistons, and valves.

working and using of the city fire engines." Of the first 28 appointed there were four blacksmiths, one blockmaker, one cutter, two gunsmiths, five carpenters, two bricklayers, one ropemaker, two carmen, four coopers, two bakers, and four cordwainers.

Following the conflagration of 1741 the Corporation bought 100 new new fire buckets and two more London engines at a cost of £100. Fourteen additional men were appointed to the engine companies, bringing the number to 44. These measures proved effective in 1742 when the firemen with four engines contained a potentially bad fire.

It appears that New York companies were the creations of the Corporation rather than the Philadelphia type of volunteer organizations formed by groups of citizens. However, like firemen elsewhere the New Yorkers received no pay. As will appear, both patterns have been followed over the years.

Unlike the working class membership of the first New York Company, those elsewhere were often organized and staffed by prominent citizens. Franklin, after praising the effectiveness of Philadelphia fire companies in preventing large conflagrations, went on to say that these companies "had become so numerous as to include most of the inhabitants who were men of property."

For instance, the Hand in Hand Fire Company boasted that it had as members the most eminent men of the city. That this was a justifiable claim is shown by a roster of eighteenth-century Philadelphia notables which included the leading physicians, the rector of Christ Church, the first Anglican bishop of Pennsylvania, several of the wealthiest merchants, the mayor of the city, and the provost of the College of Philadelphia.

The Hibernia Fire Company founded 1752, which, despite its name, was largely Protestant, had such prominent members as Nicholas Biddle (later president of the Bank of the United States), and the financier Robert Morris.

Elsewhere there was a similar pattern. George Washington was a member of the Friendship Fire Company of Alexandria, Virginia, and in 1775 presented it with an engine he had purchased in Philadelphia. In the last year of his life he was in Alexandria during a fire. Seeing the Friendship engine badly manned, he jumped off his horse and called out, "Why are you idle, gentlemen? It is your business to lead in these matters." So saying, he took hold of the engine himself, being followed by others.

In Brooklyn, where the first fire company was formed in 1785, the rules provided that men were to be chosen in the town meeting. There was an annual competition for the privilege of serving. In 1835 Fanny Kemble, writing about New York, said, "The sons of all the gentlemen here are volunteer firemen. . . ." George W. Sheldon, in his *Story of the Volunteer Fire Department of New York* (1882), stated that the roster of influential merchants and financiers who were once volunteers would make "a Homeric list." He mentioned various prominent Quaker families who were members, and among the numerous publishers: John and Fletcher Harper, William H. Appleton, James Miller, David G. Francis, and Francis Hall. In fact, Fletcher Harper was foreman of Engine Company No. 7.

In New Orleans, where the first fire company was organized in 1829, the membership during the early days was composed of leading citizens: professional men, merchants, and public officials. Later members were drawn from the best class of mechanics.

Over the years, as will appear, the composition of fire companies and the social position of their members were to fluctuate with social change in general and according to locality. In the 1850s, at a time when fire companies in cities like New York, Philadelphia, and Baltimore had become populated by rowdies, those in Milwaukee, where the first company was organized in 1837, had as members most of the important men in the city, including Harrison Ludington, later mayor and then governor of Wisconsin.

In the early 1900s one New Jersey volunteer company boasted ten millionaires.

The "articles" of the old companies are all very much alike, largely modeled on those of Franklin's Union Fire Company, which in turn were patterned after those of the Boston Fire Society. The most fundamental of the rules were those relating to a member's duty in the event of fire. The legend that companies fought fires only in buildings with special fire marks probably originated because in England fire brigades were originally sponsored by insurance companies and were expected to protect the insured's property. In America most fire companies were organized by the citizens for community service.

It is true that the articles governing the companies usually provided that members "employ our endeavours to preserve the goods and effects of such of us as shall be in danger. . . ." But the clause which followed would read, "They further agree that if their own houses are not in danger they will give our utmost assistance to such of our fellow citizens as may stand in need of it in the same manner as if they belong to this company." The Hibernia Company of Philadelphia had a similar rule but broadened it to say that members were obligated to aid others in the suburbs as well as the city.

Almost always there were requirements for the prevention of looting. The ubiquity of these rules and their detailed nature suggests that in the eighteenth century this was a serious problem. Those of the Diligent Fire Company of Philadelphia were typical: two men were to be stationed at the doors of a house to prevent suspicious persons from carrying out goods, and were to remain at their posts until everything that could be saved was packed up and taken to a safe place.

The rules always provided for salvage operations. Along with a specified number of buckets, each member had to furnish a large basket or bags for removing goods from a burning house. The Hibernia Fire Company even specified that the bags be made of at least four yards of linen of a certain quality.

There was a system of fines for failure to maintain the required number of buckets, baskets, and bags; for absence from meetings, or failure to respond to an alarm.

It is worth noting that not only did each company operate under a set of articles or bylaws, but that meetings were conducted according to parliamentary practice with regular motions to be voted on, the appointment of committees, the keeping of minutes, and regular audits of the accounts—all practices which continue to this day. A volunteer fire company, although to some extent a social organization, has always been a more structured group than, for instance, a local baseball team or the usual literary club. Thus Franklin stated that the members agreed "to meet once a month to spend a social evening" but also to discuss methods of fire fighting.

Certainly volunteer fire companies tend to outlast other local organizations or even national political parties. The Federalist, Whig, Anti-Masonic, Native American (Know-Nothing), Bull Moose, and Progessive parties are only memories—as the Republican Party founded in 1856 may well become one—but the Rainbow Fire Company of Reading, Pennsylvania organized in 1773 is still functioning. Its claim to being the oldest active volunteer unit in America may be valid. The fire department of York, Pennsylvania was founded in 1770. The part-paid company of Augusta, Maine dates from 1799; the volunteer company of Nebraska City from 1857, nine years before Nebraska became a state.

A random sampling of volunteer companies across the country showed that that of Centerville, Iowa dates from 1873; that of Tarrytown, New York from 1874; Townsend, Massachusetts, 1875; Covington, Ohio, 1880; Lake Geneva, New York, 1885; Saginaw, Michigan, 1885; Juneau, Alaska, 1895.

And as Elizabeth Smedley commented in her study of Pennsylvania fire companies, " . . . it is very unusual for a volunteer fire company to go out of existence."

III

ENJINE! ENJINE!

MEMBERSHIP IN A VOLUNTEER fire company was not a sinecure. A fireman's few compensations were exemption from jury duty, and in peace time, from military service—provisions almost universal before the Civil War. In many companies he had to pay membership dues, and, of course, fines for a variety of offenses. In New York, Engine Company No. 42, organized by businessmen, including Lorenzo Delmonico of restaurant fame, charged an initiation fee of $50. Applicants had to wait a year for admission. A member of Hose Company No. 5, when about to leave for Europe, paid $100 to have his name kept on the roll.

The 12 members of Engine Company No. 42 were known as "Apostles of Temperance" because only coffee was served. However, a member later opened a saloon next door to the firehouse. Then as now neighbors would offer food, coffee, and often liquor to men working at a fire, though not as lavishly as the dinners provided by Delmonico's uncle after a fire. However, following a response to a false alarm he set out only a cold collation.

The old New York minute books are filled with records of fines for such offenses as leaving a meeting without permission from the chairman, for profane or indecent language, for intoxication at a fire or meeting, even for smoking or chewing tobacco in the firehouse.

A member of the Hibernia Company in Philadelphia—one of the outstanding organizations—brought charges against a man for being intoxicated while wearing the ensign of membership, apparently not at a fire.

But the most common offense was absence from meetings. The minutes of the Hibernia Company have such typical entries as:

	Present	Absent	
Aug. 13, 1758	10	19	
Oct. 2, 1758	9	20	
Feb. 6, 1764	5	30	
Oct. 3, 1768	4	21	
Jan. 6, 1802	10	17	+ 2 excused
Mar. 4, 1807	12	14	
Mar. 5, 1817	9	10	

There were, of course, fines for failure to answer an alarm, and the excuses were various: In New York, Evert Duyckinck missed a fire on January 2, 1802 because there was another one three doors from his house. For January 15,

1807 "Harris Sage's Excuse is received, he says at the time of the above Fire [1 A.M.] he was lock'd in some one's Arms and could not hear the Alarm." Not accepted. E.J. White's excuse for November 2, 1807 was that he was waiting on a customer who was in a great hurry. Although he was on his way in less than 20 minutes, he was fined $1. One man in 1808 was absent from roll call because he was fishing and got caught in a strong tide. He too was fined.

On the whole the record of attendance at fires was considerably better than that at meetings, especially considering the fact that a meeting is scheduled beforehand, whereas a man may not be in the vicinity when an alarm sounds.

In the 1840s when the Hibernia Company minutes recorded the attendance at fires, the number varied between four and 45, with a usual turnout of 20 or 30 men. The minute book of New York's Lady Washington Engine Company for the 1850s shows that the number of men at a fire was between 16 and 33; that of Chatham Engine No. 27 records 11 to 41, with the usual number between 33 and 39. During the same period in Reading, Pennsylvania the Liberty Company had an average of 60 men at a fire.

Small eighteenth-century engines could be pumped by ten or a dozen men; the much larger ones of the 1840s, '50s, and '60s required 20 or 30. A Philadelphia-type engine built by John Agnew could, because of its two tiers of brakes, be manned by 48 men. As has been noted in Chapter II, bystanders could be pressed into service at the brakes. And in the nineteenth century many city companies had an unofficial group of "runners," chiefly teen-agers who would run ahead of the engines, do menial chores, or even be pressed into service at the pumps. These auxiliary groups were sometimes organized with their own officers, and with fines for nonperformance of duties.

Because the first engines had no hose but depended upon a

pipe, called a gooseneck, mounted on the machine, men had to get very close to a fire as is shown in old engravings. Ordinary citizens, men, women, and children formed bucket lines to fill the tub on the engine.

The large number of conflagrations involving numerous buildings was due to a variety of causes such as the prevalence of frame buildings, the mechanical limitations of the early engines, and especially to lack of adequate water supply. Boston was notable for being honeycombed with crowded alleys, many of which were wide enough only for pedestrians.

Before the advent of piped water firemen had to depend on wells, cisterns, ponds, and streams. A man in Boston got the idea of keeping 100 gallons of water in tubs, a practice copied by other companies, but the water froze in unheated engine houses, and the practice was forbidden. Although engines with suction hose had appeared in the 1730s they seem not to have been widely used until the nineteenth century. When suction engines were finally adopted, they did away with the need for a bucket line, and led some municipalities to install cisterns strategically placed for fire protection. The introduction of leather hose had made it possible to keep the engine at a distance from the fire and to permit engines to be linked in relays. Hose also made it possible for firemen to work inside buildings but with the hazards of smoke inhalation and falling walls.

At first an engine carried one length of hose, until 1797 when Mr. John Halsey of New York donated several lengths imported from Hamburg. In 1803 the Philadelphia Hose Company obtained 600 feet of 2¼-inch leather hose at $.43 a foot. Eleven sections of this were in 50-foot lengths (the usual length of sections today) and two were each 20-feet long.

The introduction of hose of considerable length was responsible for the organization of hose companies independent of the engine companies. At first they carried the hose in a box on wheels but soon began to use a vehicle with a reel. The use of hose introduced a new kind of drudgery for firemen: after use leather hose had to be cleaned, dried out, and oiled. In fact, until the appearance of nylon hose in the 1960s the unrolling and drying of hose was a universal chore.

Usually the engine company did the actual fire fighting. The first man to answer an alarm had the privilege of playing the hose pipe. The pipe of No. 38 in New York was so heavy it required three men. (A modern 2½-inch hose line requires at least two fire fighters at the nozzle.)

Hazards were not confined to men on the hose line: if a man pulling the apparatus slipped and fell, he had to roll out of the way quickly to avoid being crushed or suffer a broken limb. Some fatalities from this cause were recorded. As young Henry James learned, a man could be hurt while operating the brakes (which were the long handles pumped by the men). At the normal rate of 60 strokes a minute, a man would be exhausted after 10 minutes of pumping; when the count went up to 120 strokes, he had to drop out after 3 minutes. Men relieving others at the brakes constantly suffered crushed fingers, broken arms, or heads.

In the 1880s W.L. Jenkins, president of the Bank of America and a former member of New York Engine Company 13, remembered pumping his breath out, lying in the street to recover, then jumping up to pump again. Another old-time New York volunteer recalled that "Firemen were continually over-exerting themselves by overwork. . . . We were burying men all the time who died from the same cause. . . ."

In the cities, firemen were kept constantly busy. European visitors were amazed at the number of fires in New York. The often acidulous Capt. Basil Hall, who traveled in the states during 1827 and 1828, wrote of "that city which seems to be more plagued with fires than any town in the world. . . ." Fanny Kemble under the date September 16, 1832 noted, "There are on an average, half a dozen fires in various parts of the town every night." Writing of Philadelphia, Charles L. Yell, whose travels cover the years 1841-1842 stated, "We were five days here, and every night there was an alarm of fire, usually a false one. . . ."

What this meant to the firemen is revealed in the minutes of the companies. For instance, in 1840 the Hibernia Company answered six alarms in November and 17 in December, three of them on Christmas morning. In October of 1842 they answered 14 alarms, but on two of them did not go into service, which probably meant that other companies had the fires under control. Men often did not get much sleep: one day there were fires at 1 A.M. and 7 A.M.; on another 5 A.M. and 6 P.M.

These figures are not unlike those for a modern volunteer company in a heavily populated area. For instance, in April 1976 the Tarrytown, New York Fire Department with six companies and 500 volunteers answered 75 alarms. The Reading, Pennsylvania Fire Department with 15 companies answers over 1,000 calls a year. The seven volunteer companies in Lower Merion Township and the Borough of Narberth on Philadelphia's Main Line answered 1,048 alarms in 1974 and 1,095 in 1975.

The Hamptown Township Volunteer Fire Department in Massachusetts has about 150 calls a year, but the ambulance service averages 600 to 650 annually. In Alaska the Ketchikan Volunteer Fire Department, located on a small island and with 11 full-time men and 45 volunteers, has approx-

imately 250 annual fire calls and about 1,000 ambulance
and rescue calls. On the other hand, a small company in
Eleanor, West Virginia had only 45 runs in 1975. In Oregon,
where the State Fire Marshal's office publishes statistics,
some rural companies have only one to five calls a year;
whereas the predominantly volunteer companies of Forest
Grove had 362; Illinois Valley, 255; Monmouth, 299; and
Woodburn, 375.

One difference between the present and the past is that
calls were once almost exclusively for fires or major calami-
ties; today they may be for a child locked in the bathroom,
for smoke odor, for faulty electrical appliances, for an over-
heated pan on the stove, or even for a cat in a tree. A
hundred and fifty years ago, a householder with a minor
crisis called in a neighbor; today he or she picks up a phone
and calls the fire department. The old lady who is constantly
smelling smoke is a ubiquitous type.

A great difference between the fire calls of the past and
those today is that until about the 1860s in the cities and
1920s in villages the firemen pulled their engines by hand.
By the 1840s Pat Lyon's Philadelphia-type of double-decker
end-stroke engine, many of them built by John Agnew, became popular in Boston, Manhattan, Brooklyn, and elsewhere. Weighing 4,000 to 4,800 pounds and with two tiers
of brakes, it could be pumped by 30 or as many as 48 men,
one crew on the ground and another on platforms front and
rear.

In the old days a man might work to exhaustion on one
fire only to be called out for another a few hours later. This
is not to belittle the demands upon the modern firefighter:
dragging hose up ladders, standing on a sloping roof while
chopping through slate shingles, rolling up perhaps 2,000 or
even 4,000 feet of hose—at least an hour's work—are all no
picnic. Furthermore, the size of modern buildings, such as

apartment houses with many units, requires today's fire fighter to work in breathing apparatus consisting of a mask and
a 40 pound back cylinder of air. Every climb on a ladder or
hose carried up a stairway becomes a double chore.

In Europe the famed prowess of American firemen, especially those of New York, led visitors to America to observe them in action. Baron Klinkowström, who traveled
here from 1818-1820, commented on the prevalence of fires
in New York, but added that it seldom happened that more
than one wooden house burned down:

> The fire departments are excellent: the water supply is
> nearby; and discipline is good even without a garrison.
> Every district in the city is furnished with a fire engine,
> and owners of houses and businessmen who live in the
> section are compelled by law to serve. Some handle en-
> gines, others keep order. . . . Often people of the better
> classes dress in coarse firemen's outfits with big helmets
> with a wide brim, and large leather aprons to do menial
> labor. The clothes are always kept in good order in the fire
> houses as are the equipment and the tools.
>
> Usually horses are not hitched to the fire engines, but
> the firemen and all who pass by grab the shafts and run to
> the fire. . . . Once I was standing idly looking at the prepa-
> rations and suddenly I was required to join the crowd,
> take hold of the shafts, and pull the engine to the fire.

He thought that it would be fine if the European public
could learn to keep order without a garrison during
emergencies.

Perhaps the best account of American fire fighting in the
1830s is that by Capt. Basil Hall. No great admirer of America, he had, nevertheless, heard of the skill of New York
firemen. So anxious was he to see them in action that hear-

ing an alarm at two in the morning, he got up and dressed. To his disgust, it was a false alarm and the engines were returning as he got to the door. But scarcely had he got back to sleep when there was a far more furious alarm. This was the real thing:

I succeeded by quick running in getting abreast of a fire engine; but although it was a very ponderous affair, it was dragged along so smartly by its crew of some six-and-twenty men, aided by a whole legion of boys, all bawling as loud as they could, that I found it difficult to keep up with them.

Four houses, built entirely of wood, were on fire from top to bottom, and sending up a flame that would have defied a thousand engines. But nothing could exceed the dauntless spirit with which the attempt was made. In the midst of prodigious noise and confusion, the engines were placed along the streets in a line, at a distance of about two hundred feet from one another, and reaching to the bank of the East River. . . . The suction hose of the last engine in the line, or that next the stream, being plunged into the river, the water was drawn up and then forced along a leathern hose or pipe to the next engine, and so on, till the tenth link in this curious chain, it came within range of the fire. As more engines arrived they were marshalled by the superintendent into a new string; and in about five minutes after the first stream of water had been brought to bear on the flames, another was sucked along in like manner, and found its way, leap by leap, to the seat of the mischief.

. . . On retiring reluctantly from this interesting scene, I caught a glimpse of a third jet of water playing around the back part of the fire; and on going round to that quarter, discovered that these energetic people had formed a third series, consisting of seven engines, reaching to a different bend of the river. . . .

The chief faults he found were the needless shouts and uproarious noises which tended to exhaust the men at the engines, and the fool hardiness with which men entered burning buildings and climbed up on them by means of ladders.

Other observers of the American scene commented upon the uproar at fires. Writing about New York firemen in 1832 Fanny Kemble noted, " . . . the great delight they take in tearing up and down the streets, accompanied by red lights, speaking trumpets, and a rushing, roaring escort of running amateur extinguishers, who make night hideous with their bawling and bellowing."

The traveler Charles L. Yell described the Philadelphia scene in the early 1840s:

> . . . the noise of the firemen was tremendous. At the head of the procession came a runner blowing a horn with a deep unearthly sound, next a long team of men . . . drawing a strong rope to which the ponderous engine was attached with a large bell at the top, ringing all the way, next followed a mob, some with torches, others shouting loudly; and before they were out of hearing, another engine follows with a like escort; the whole affair resembling a scene from *Die Freischutz* or *Robert le Diable,* rather than an act of real life.

However, he granted that all this was no sham: these young men were ready to risk their lives, "and as an apology for their disturbing the peace . . . we were told that the youth here require excitement!" He thought they managed things as effectively in Boston without turmoil.

A veteran of the New York Fire Department remembered that the famous Zophar Mills * "had a throat like a lion's." As a young man sleeping in an attic so that he could hear alarms, the voice of Mills came to him from five blocks away: "Turn out! Turn out! Fire! Fire!"

The tremendous efforts of volunteer firemen were perhaps best exemplified during the great fire of 1835 in New York. About 9 P.M. of December 16, with the temperature at 17° below zero, a watchman saw a fire in a five-story building. The ringing of church bells brought an engine company to the scene in ten minutes. Chief "Handsome" Jim Guilick ordered a general alarm, but by 10 o'clock 40 buildings were ablaze. Firemen lowered engines to the river and chopped holes in the ice, but water often froze in the hoses. Because of freezing spray from those nozzles which worked, men cut head holes in blankets and put them over their coats. Engine No. 33 was run out on the deck of a brig to take suction to feed other engines. The cook made a fire in the galley where six men at a time would warm up, then someone would put a hat over the stovepipe to drive out the men and admit a new shift. The engine kept going all night, but during a pause the next morning, it froze up.

The mayor sent to Governor's Island for sailors and marines with powder to blow up buildings. Chief Guilick put the legendary Zophar Mills and his Eagle Engine No. 13 at the narrowest part of Wall Street where they stopped the fire before it devastated all of lower Manhattan. Altogether 654 buildings were ruined.

During the conflagration firemen came by ship, rail, and on foot from Brooklyn, Jersey City, Newark, White Plains, Hoboken, and Morristown, but the saga of the 400 firemen

---

* Zophar Mills, a fireman for 45 years, was described by a man who remembered him as "the crème de la crème of a fireman." Sheldon, p. 47.

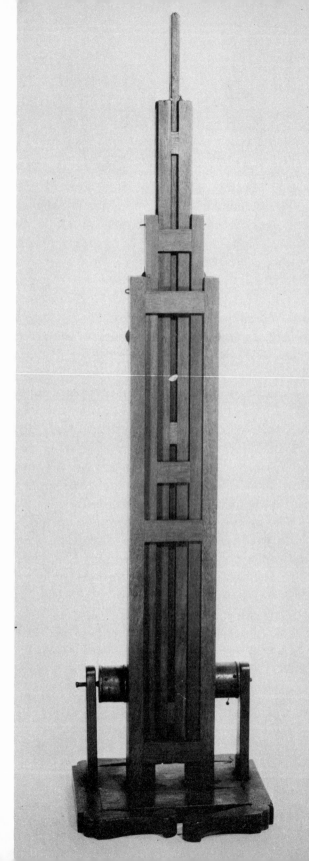

The Speedy
Elevator, invented by
Nicholas Collin, D.D., in
1791 (*Courtesy American
Philosophical Society*)

Ladder truck built for the Philadelphia Fire Company in 1799 *(Courtesy INA Corporation Museum)*

The first hose carriage in the United States, used in Philadelphia in 1803. It was equipped with bell, nozzles, and glass boxes for candle lights. It was hand drawn and painted green. The original was built by Patrick Lyon. The model is in the INA Museum. *(Courtesy INA Corporation Museum)*

Model of Philadelphia-type engine, double deck, end-stroke, of the Fairmount Company of Philadelphia, founded 1823. *(Courtesy INA Corporation Museum)*

Panel of the Franklin Engine Company of Philadelphia, painted about 1830 *(Courtesy INA Corporation Museum)*

Left: Panel from the Washington Engine Company. Engine panels, placed on the side of engines, were a source of pride and prestige for the nineteenth-century volunteer fire departments.

Below: The Great Fire of New York, 1835, as seen from the top of the Bank of America, corner of Wall and William Streets. Painted by Nicolino Calyo, engraved by W. J. Bennett. *(Courtesy of the New York Historical Society, New York City)*

of Philadelphia is the most remarkable. In response to an appeal from the mayor of New York the men set out with the machine of the Franklin Engine Company. Twenty-three pulled it to the Walnut Street wharf where they were joined by other firemen. Because of ice, the ferry could not cross to Camden. The men then started by train for Kensington but were stopped by ice on the rails. Starting for Trenton on foot they reached it at midnight. In a tavern there the ice on their coats and hats melted, flooding the floor.

On the road again they hired a man with a horse to pull the engine. At a house where they offered to pay for supper the owner refused to serve them, but seeing food on the table they went in and ate it. At the Sand Hills only six men were left with the engine plus the man with the horse. After a night's rest in a tavern they were joined on Sunday by the rest of the firemen. By Monday the rails were cleared of ice and a locomotive with two flat cars came along. Putting the front wheels of the engine on one car and the rear wheels on another they got it to Amboy where a steamer took them to New York.

By then the fire was out, but the mayor arranged to have them put up at the Astor House where 15 gentlemen, including the mayor and John Jacob Astor, gave them all kinds of refreshments. The next day Fire Company No. 1 came to visit and take them on a tour of the ruins. The Philadelphians were invited to the theater, but refused to go unless the men of No. 1 were also invited. They were. The narrator, J.B. Harrison of the Franklin Company, found that when he went shopping, a merchant refused to accept money. When the Philadelphia men were ready to go home, New York firemen pulled their engine to the wharf, and a committee appointed by the mayor escorted them home. From that time on, firemen began visiting back and forth between the two cities.

In a book published in 1851, Lady Emmeline Stuart Motley wrote of America:

Nowhere on the earth, I should think, are such numerous and splendid bodies of firemen; and in no place under the sun or moon, I honestly think, have they such extensive and incessant and unlimited practice. And what men in the world ought to make such admirable warriors as firemen? At all times, but especially at the dead hour of midnight, forced to leave their homes at a moment's notice, to start from slumber, after perhaps a day of wearying toil and harassing exertions—to confront the direst extremes of cold and heat—to face the raging element, that is their remorseless and tremendous antagonist—to dare every imaginable peril without prospect of reward, or of promotion, or even renown and glory—they should certainly make heroes when fame and victory beckon them away.

There are, of course, tales of narrow escapes and heroic rescues. In the New York conflagration of 1845, the blaze started in a sperm oil store, raged for 12 hours and destroyed 200 buildings. One of them was a warehouse with a large amount of saltpeter which exploded, shattering a million pieces of glass and rocking Manhattan, Brooklyn, and Jersey City. The blast lifted the roof, which landed intact on a building across Broad Street, and carried with it fireman Francis Hart, who suffered a dislocated ankle.

An arsonist in Baltimore, April 14, 1857, threw a pail of blazing coal oil into a store, causing a conflagration that involved all the fire companies in the city. Frank Welsh of the First Baltimore Hose Company went home exhausted, went to bed, and, dreaming that he was still fighting the fire, walked out of a second floor window in his sleep. He was knocked senseless but recovered.

In New York when the Park Theatre burned in 1848, Malachi Fallon dashed in and carried out the beautiful Miss Dyott, an actress. He was one of the firemen who later left for the gold rush. In the West he distinguished himself by helping to catch a suspected murderer, and became San Francisco's first chief of police.

When Barnum's museum burned in 1865 a huge Bengal tiger got loose. Fireman Johnny Denham felled it with an axe, then carried out the fat lady weighing 400 pounds. Not content with that feat he next rescued two children and the woolly-headed albino woman.

The "Fireman save my child!" gag could get a laugh, but it was based on sober reality. It could produce unexpected results: at a Milwaukee fire in 1870 the chief and another man climbed up a ladder in response to a woman screaming, "My baby!" The chief brought the woman down and the other man hurled to the street a bundle she had been carrying. When men ran to pick up the mangled infant, they found a badly frightened canary.

Now as then a fire fighter's first responsbility is to save human life. Over and over he is told at drills, "Get the people out first." Today volunteer fire fighters don masks and air canisters, then go into dark rooms to practice searching under furniture and in closets—anywhere a fightened child might hide. Fire fighters are told to feel under beds for unconscious victims.

Fire fighting has always been hazardous, but was especially so in the nineteenth century when buildings had become large but were wall-bearing. Thus when beams burned, the masonry walls tended to come down. Sealed breathing apparatus had not been developed. In one instance when a falling wall engulfed some New York firemen, Chief

Engineer James Guilick let the water out of an engine and used it to pump air into the ruins. At the same fire Zophar Mills fell two floors.

One of the most vivid accounts of volunteers in action appeared in *All the Year Round,* March 16, 1861, possibly written by Charles Dickens. The mixture of vivid writing, melodrama, and sentimentality suggests his work.

But, after all, it is in night-time that the fireman is really himself, and means something. He lays down the worn-out pen, and shuts up the red-lined ledger. He hurries home . . . slips on his red shirt and black dress-trousers, and dons his solid japanned leather helmet bound with brass, and hurries to the guard room, or station if he is to be on duty.

A gleam of red, just a blush in the sky, eastward . . . and presently the telegraph begins to work. For every station has its telegraph, and every street its line of wires, like metallic washing lines. Jig-jag, tat-tat, goes the indicator. . . .

Presently the enormous bell, slung for the purpose in a wooden shed in the City Park, just at the end of Broadway, begins to swing and roll backward.

In dash the volunteers in their red shirts and helmets— from oyster cellars and half-finished clam soup, from newly begun games of billiards, from the theatre, from Boucicault, from Booth, from the mad drollery of the Christie minstrels, from stiff quadrille parties, from gin slings, from bar-rooms, from sulphurous pistol galleries, from studios, from dissecting rooms, from half-shuttered shops, from conversation and lectures—from every-where—north, south, east, and west—breathless, hot, eager, daring, shouting, mad. Open fly the folding doors, out glides the new engine—the special pride of the company—the engine whose excellence many lives have been lost to maintain. . . .

Now, the supernumeraries—the haulers and draggers, who lend a hand at the ropes—pour in from the neighboring dram-shops or low dancing-rooms, where they remain waiting to earn some dimes by such casualties. A shout—a tiger!

"Hei! Hei! Hei!" (cresendo), and out at lightning speed dashes the engine, in the direction of the red gleam now widening and sending up the fan-like radiance of a volcano.

Now, a roar and crackle, as the quick-tongued flames leap out, red and eager, or lick the black blistered beams—now, hot belches of smoke from shivering windows—now, snaps and smashes of red-hot beams, as the floors fall in—now, down burning stairs, like frightened martyrs running from the stake, rush poor women and children in white trailing night-gowns—now, the mob, like a great exulting many-headed monster, shouts with delight and sympathy—now, race up the fire-engines, the men defying each other in rivalry, as they plant ladders and fire escapes. The fire-trumpets roar out stentorian orders—the red shirts fall into line—rock, rock, go the steel bars that force up the water—up leap the men with hooks and axes—crash, crash, lop, chop, go the axes at partitions, where the fire smoulders. Now, spurt up in fluid arches, the blue white jets of water, that hiss and splash, and blacken out the spasms of fire, and as every new engine dashes up, the thousands of upturned faces turn to some new shade of reflected crimson, and the half-broken beams give way at the thunder of their cheers.

The fire lowers, and is all but subdued, though every now and then a floor gives way with an earthquake crack, and into the still lurid dark rises a storm of sparks like a hurricane of fire-flies. But suddenly there is a crowding together and whispering of helmeted heads. Brave Seth Johnson is missing; all the hook men and axe men are back but he; all the pumpers are there, and all the loafers are there. He alone is missing.

Caleb Fisher saw him last, shouts the captain to the eager red faces; he was then breaking a third floor back window with his axe. He thinks he is under the last wall that fell. Is there a lad there will not risk his life for Seth? No! or he would be no American, I dare swear.

Hei! hei!! hei!!! hei!!!!

Click—shough go the shovels, click-click the pickaxes. A shout, a scream of "Seth!"

He is there, pale and silent, with heaving chest, his breast-bone smashed in, a cold dew oozing from his forehead. Now they bear him to the roaring multitude, their eyes aching and watering with the suffocating gusts of smoke. They lay him pale, in his red shirt, amid the hushed voiceless men in the bruised and scorched helmets. The grave doctor breaks through the crowd. He stoops and feels Seth's pulse. All eyes turn to him. He shakes his head, and makes no other answer. Then the young men take off their helmets and bear home Seth, and some weep because of his betrothed, and the young men think of her.

Such are the scenes that occur nightly in New York.

IV

THE LIFE OF A FIREMAN

"For the last six nights I've been trying to
discharge my conjugal responsibilities," one
fireman said, "and every time I get started
that damn bull horn. . . ."
JOHN CHEEVER, the *Wapshot Chronicle*

To AN AMAZING DEGREE a man's service as a fireman was the
center of his emotional life. He might earn his living as a
mechanic or a banker; he might have a wife and children,
but he was first of all a fireman. A volunteer born in 1799
said, "A fireman thought more of his engine than of his
family." As fire historian Robert S. Holzman commented,
"He probably saw more of it." Zophar Mills recalled that
there was a fire on his wedding night: "I could see it, and I
wanted to go immediately. But the next morning early—
before breakfast, there was another fire, and I went to that."

A famous member of the Old Turk Company, Frank Clark,
joined as a runner at the age of 11. When the foreman
offered a new suit to the first runner to answer the next

48

alarm, the men answering one at midnight found Clark waiting at the tongue of the engine—stark naked. He became foreman himself in 1847. A year later, going home from his wedding, he was near the door when the firebell rang. Abandoning his wife he went to the fire in his wedding suit. She had no key to the house, so went home to her parents. He did not see her for three days.*

Of course, wives sometimes had something to say about all this. The minute book of New York Engine Company 21 under the date June 8, 1811 records: "Held an election for a member in place of L. Chapman, resigned; the reason was his wife was afraid he would burn himself."

In the 1880s an old-time former fireman bemoaned the days when women took their places in the bucket line:

Alas for the degeneracy of latter years, when the tinsel and frippery of fashion has driven out of use those comfortable and numerous garments familiarly known as 'linsey woolsey petticoats,' sufficient in fabric and quantity to protect from the effects of water the female fireman. Changed indeed is the aspect of the voluntary aid system in this respect; no *gude vrouw* is found enlisted in it, but instead of these services of the brawny-armed fair sex, the fireman having toiled until he had overcome the devouring element, wends his weary way home, uncheered by the refreshing thought that the gentler sex have shared in his labors, and only perhaps to go to sleep under the soothing influence of a curtain lecture on the impropriety of venturing his health and life and consequently the happiness of his family by endeavoring to serve others.

*The Townsend, Mass. Fire Department cherishes a story of a wife waking her husband to tell him it was time to take her to the maternity hospital. "Wait a minute," says Dan, "the fire whistle's blowing."

As New York grew larger and men did not necessarily live near the engine house, the unmarried men tended to set up dormitories nearby. Out of their own pockets men rented bunk rooms. Engine Company 33 got the sexton of All Saints Episcopal Chapel to let them sleep in the pews. When they were ejected, they rented a room, then moved to a building where 20 men slept every night. At the foot of their bunks the men placed their boots with the trouser legs in place over them. When an alarm sounded, a man had only to step into his boots, pull up his pants, and run downstairs for his fire coat. An old member of Engine Company 14 telling of his younger days said, "I slept in a bunk room six years on a straw mattress, on a common cot. . . ." Some dormitories became elaborate with lace pillows, fine carpets, drapes, ornate chandeliers, and oil paintings. A number of them contained libraries. In Reading, Pennsylvania the Junior Fire Company started a library in 1856, and the Liberty in 1867. The latter eventually contained the *Britannica*, books on government and history, fiction by Cooper, Holmes, Dickens, Dumas, Kipling, and O'Henry.

As in a college fraternity house, there was horseplay. One joke was to take lampblack and paint moustaches on sleeping men. In some companies the bunk rooms were kept orderly with a 10 P.M. curfew. Not all companies permitted bunking because of the likelihood of drinking and carousing—a scruple not characteristic of colleges. However, after the introduction of steam fire engines it became necessary to have at least some men sleep in the firehouses in order to take care of the horses and keep the buildings heated, for steam boilers could freeze.

But whether or not a man slept in a bunk room, he had to be ready at all times. In his old age Peter Warner told of men who would throw down their tools and lose half a day's or even a day's pay, who would ruin clothes and have to buy new ones. Charles Forrester, another old-time fireman, boasted, "I never lost a day in my business. Often I was with my engine company four nights a week, yet I was at work as usual in the morning. . . . The excitement kept us up, I suppose." One night he was at home taking a vapor bath for a cold when the bell rang. He threw off his blankets and, dripping wet with perspiration, ran with the engine. "The next morning I never felt better."

It all still happens 150 years later: the loss of pay, the ruined clothes, the jumping out of baths and showers. Suburban firemen have turned out in dinner jackets. In Gladwyne, Pennsylvania a man who had put on his turnout gear over a nightshirt yelled all the way to the fire because of the cold updraft on the rear step plate of the engine.

As Foster commented, "We were fully repaid when we could brag about our exploits, and make our neighbors feel jealous." The firemen interviewed by George W. Sheldon in the 1800s—40 or 50 years after their time of service—all told the same story. "I would rather have lost a dozen teeth than resign from my company," said Warner. Ex-Park Commissioner James F. Wenman recalled his fire service as "the happiest days I ever spent in my life." Carlisle Norwood, president of the Lorillard Fire Insurance Company, and formerly foreman of Hose Company No. 5, said, "I thought there was nothing like being a fireman. I would sooner go to a fire than to a theater or any other place of amusement. There was no pleasure that equaled that."

A Mr. George T. Hope said, "When I look back upon my experience as a fireman, I think of it in this wise: I was of use as a citizen; I made friendships that have never been broken; and I gained a knowledge of my own business as an insurance officer. . . ." He said he had attended thousands of

fires, as many as 60 in 60 days. As another veteran re-
marked, "If a fellow took a fancy to going to fires, you might
as well kill him as try to stop him."

It was considered a great disgrace to have one's engine
"washed," that is, caused to overflow by the pumper behind
it in line. The story is told of a fireman visiting a consumptive companion in the old New York House of Refuge and
mourning, "Oh Jake, could I but be in your place at this
moment, it would be a happiness to what I now suffer. . . .
Jake, the engine got washed today."

"Dick, who washed her?"

"Twelve engine."

"Then let me die, for I envy not your hold on life." The
consumptive promptly did die. True or apocryphal, such
stories capture the spirit of the old fire companies.

When New York Engine Company No. 13 got a new
pumper in 1835, they offered a new suit to the foreman of
the first company to wash them. Company No. 11 won by
pumping 130 to 150 strokes a minute. A man could last only
15 seconds at the brakes.

In Brooklyn, Protection No. 6 with a piano box engine was
always victorious in washing contests. They were known as
the "Bean Soup Company" because a Mrs. Boyd had fed
them bean soup after a fire. When Company No. 1 got a new
engine, they issued a challenge stipulating that each side put
up $500 to test which could wash the other in five minutes.
Again the Bean Soup Company won. On a less happy occasion in 1853 Engine Companies 2 and 7 held a washing
contest. After No. 7 won, it was found that the valve on No.
2 had been "hung up," perhaps by sabotage. This led to such
bitter enmity that at fires one or the other company was
prevented from throwing a stream.

Old-time firemen were fond of recalling with pride the
times their engines had behaved themselves handsomely. A

minute book recorded under December 14, 1808 the extinguishing of a fire in buildings three- and four-stories
high: "No. 13 did her part that night." The secretary of New
York Engine Company No. 21 noted: "Thomas Franklin and
B. Strong were present at the playing of the engine and
pronounced her the best one in the city." And on December
15, 1825, "Our engine worked unusually well this morning,
and the members were in fine glee." After a fire men would
sometimes kiss an engine which had performed well.

All this could be echoed by any fireman in the 1970s,
especially if his engine outperformed that of some other
company. Men will still talk fondly about some pumper long
since traded in. Occasionally men will raise money to keep a
favorite engine from being sold when a new one has been
purchased.

In the past as well as now men who had fought a fire were
offered alcoholic drinks. As one song put it:

And when the fire was mostly quenched
　　And smoke obscured the stars,
Some trump with open heart would treat
　　To lager and cigars.

A New York company minute book under the date August
12, 1813 recorded: "Mr. A. Wright adjacent to the fire offered us on duty a good supply of Brandy & Water & at the
conclusion a good breakfast with hot coffee, retired in good
order & sober."

About 1800 the office of steward was established. At a fire
he had a small keg of Hollands gin and bites of bread and
cheese. Some companies attached a keg to the shaft of their
machines. Old-time firemen would sometimes pour brandy
into their boots to keep their feet from freezing.

Hope engine of Philadelphia built by John Agnew in 1838. Because of the brakes at the end of a Philadelphia-style engine, it could be operated in a narrow street. *(Courtesy of the Firefighting Museum of The Home Insurance Company)*

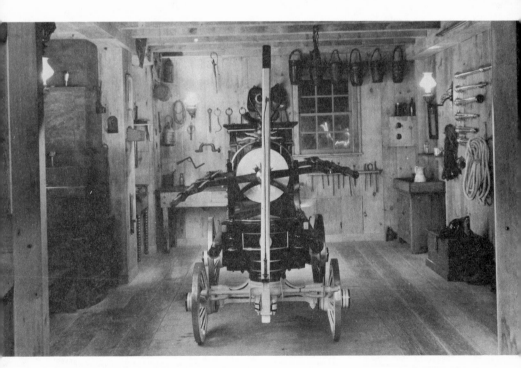

Reconstructed firehouse of Eagle Engine Company No. 13. The original was located on Maiden Lane, New York City. The setting is early to mid-nineteenth century. *(Courtesy of the Firefighting Museum of The Home Insurance Company)*

The Burning of Pennsylvania Hall, Philadelphia, 1838. Mezzotint by John Sartain from a drawing made on the spot by the artist. The Hall, built by abolitionists, was fired by a mob at night. The anti-abolitionist sentiment of the firemen is shown by their refusal to fight the blaze in the Hall; instead they tried to save adjoining buildings. *(Courtesy Library Company of Philadelphia)*

Signal lamp of the Essex Company, Salem, Massachusetts, about 1840 *(Courtesy INA Corporation Museum)*

144 VOLUNTEER FIRE DEPARTMENT OF THE CITY OF NEW YORK.

Engine Company No. 10 once received a banner under circumstances fully described by the following handbill, which was circulated at the time:

Grand Presentation
OF A
BANNER,
BY THE
LADY LAFAYETTE
Temperance Benevolent Society,
TO
FIRE ENGINE Co. No. 10.
On Tuesday Evening, June 28, 1842,
IN THE
CHURCH CORNER DELANCEY AND CHRISTIE STREETS.

PART I.	PART II.
PRAYER, ... By Rev. Mr. MARTIN.	ADDRESS, ... Mr. SNOW.
	Alcohol is Going . . . . Green.
Temperance Millenium . . . . . L. L. T. B. Society.	Engine and Tight hose . . . . . Master Madden.
Welcome Lafayette . . . . Bogart.	Song . . . . Nagle.
Old Oaken Bucket . . . . Madden.	ADDRESS, ... Mr. LATHAM.
Song . . . . Duncan.	Ode—"Temperance Ship and Temperance) Murphy.
	Crew." Words by J. Aikman . . . . . .)
ADDRESS, ... Mr. BRUSH.	Trumpet Song . . . . Bogart.
	Song—" Long Lost Youth." Words by Mrs.) Lebarn.
Air, Hail Columbia . . . . Brass Band.	Depu . . . .)
Ode . . . . Fire Co. No. 10.	Ode . . . . Master Brown.
Song—"Temperance Tree" . . . . Lebarn.	Song . . . . Duncan.
Duett—"Drunkard's Resolve" . . Murphy and Lebarn.	Ode . . . . Fire Co. No. 41.
Song . . . . Nagle.	ADDRESS, ... REV. MR. MARTIN.
Presentation of Banner . . By a Lady of L. T. B. Society.	Ode . . . . Madden.
Reception of Banner . . Foreman Fire Co. No. 10.	Song . . . . No. 15 Hose Co.
Air . . . . Brass Band.	Sister's Call . . . . Three Ladies.
	Ode . . . . Fire Co. No. 58.
	Ode . . . . Ladies' L.T.B. Society.
BENEDICTION, ... REV. C. N. HAWLEY.	

N. B.—JAMES GULICK, Esq., and Mr. C. V. ANDERSON, Chief Engineer, are expected to be present on the occasion.

TICKETS ONE SHILLING each, to be obtained at No. 335 Bowery, 364 Bowery, 267 Bowery, office of the *Washingtonian*, and at the door on the evening of the presentation.

Oliver, Cheap Cash Printer, "Organ" office, cor. Ann and Nassau-sts.

A handbill announcing the presentation of a banner to Engine Company No. 10.

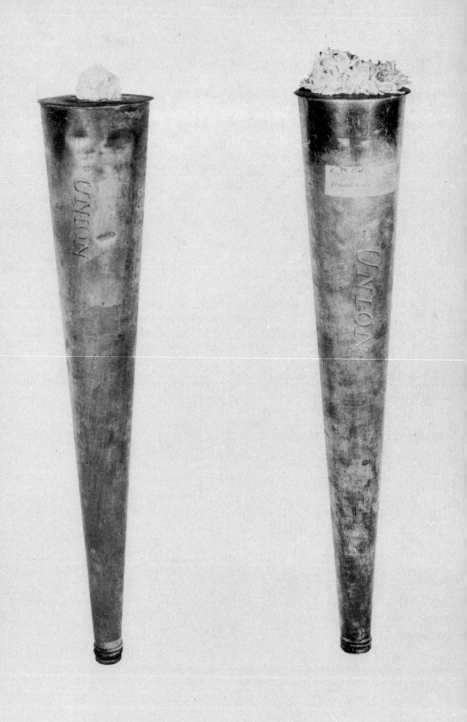

Flambeaux, tin containers packed with tallow and topped by bunched waxed wicks, carried by hand to guide engines and hose reels along dark streets. *(Courtesy INA Corporation Museum)*

Tomb of Jefferson Fire Company No. 22 in 1852. Lafayette Cemetery, New Orleans, Louisiana. (*Courtesy Leonard V. Huber,* New Orleans, a Pictorial History)

This cartoon drawn by Nokes and published 1853 by Peter E. Abel was in response to a committee report, January, 1853, recommending a paid fire department in Philadelphia. At the left volunteer firemen in traditional uniforms fight the blaze, whereas the group with the double-decker engine labeled "Paid Fire Department" are shown in great confusion. Two men standing by an overturned engine say it has just killed a woman. Another engine has broken two wheels and the horses are bolting. By contrast to the balky horses of the paid men, the volunteers are pulling their own apparatus. Two policemen escort a paid man caught looting. He gives low pay as an excuse. Little is known of either Nokes or Abel, but the latter was a bookseller. It is probable that both were volunteer firemen. (*Courtesy of the Mutual Assurance Company*)

In May 1837 after an all-night fire George Templeton Strong noted that about daylight, when the danger was about over, "the firemen of No. 17 eased off a little and attacked some provender, eatable and drinkable, ravenously and no wonder; they had worked desperately and they drank water by the gallon. Some of them refused to drink anything else." However, some of the men drank ale and wine "on which they got somewhat cocked, and rather boisterous, but I'm sure they were excusable after such a night's work, and a more orderly, considerate, well-bred set of fellows no one could desire."

The nineteenth century was a hard-drinking era, and firemen were under special temptation as Strong's account makes clear. It was also a time when the temperance movement was active. (By *temperance* its advocates meant total abstinence from all alcohol, including beer and wine, though some white-ribboners used highly alcoholic tonics like Peruna and Lydia Pinkham's Compound.)

Obviously, some of the firemen observed by Strong were teetotalers. In the 1830s and '40s Temperance Benevolent Societies, mostly led by ladies of some church near a firehouse would hold a Grand Presentation of a banner to a fire company which took the pledge. In Rochester, New York a crusade among firemen prompted 16 to take the pledge in 1843. Arrangements were made to serve hot coffee instead of brandy at fires.

In view of the growth of drunken brawls and rowdyism in city fire departments (to be discussed later) it is doubtful if the temperance movement was widely effective.

The fire company was the center of the social and leisure-time activities of its members. Even today in small communities it rivals the local Legion Post in that respect. Sunday was a favorite time for shining up the engine and putting gear in order. In Reading in the 1870s firemen started attending church in a body. But the leading activities were balls, banquets, picnics, and parades.

The housing of a new engine was and is an occasion for a ceremony and feed to which other companies and sometimes ladies are invited. An early ritual still observed in some places is to take the new engine out for a demonstration of its effectiveness. An old custom that continues in some companies is for the men to push the new piece of apparatus into the firehouse three times. (Nowadays this requires an unobtrusive assist from the motor.) Other companies ceremoniously wash off the engine, a kind of baptism. In 1829 the secretary of New York No. 13 recorded that during the day of a housing hundreds of people had visited. "As regards to her Painting, Gilding, Plating, & Carving, she never will probably be equalled."

In New York, Philadelphia, and elsewhere, firemen's balls were a regular feature. For the anniversary of the formation of the Philadelphia Fire Association the firemen held a parade on March 27, 1824 with 40 engine and hose companies, followed by a ball at the Municipal Fund Hall. The Reading Hose Company No. 1 held their first Grand Ball in 1848 and six years later the Rainbow Company began the custom of an annual ball.

In 1856 Professor Henry G. Thunder of Philadelphia composed "The Empire Hook and Ladder Polka." A composition described as less delicate was "The Quadrille Fire Set," orchestrated for various instruments including a fire bell. Then there was the danceable "Fireman's Bride:"

Who wouldn't be a fireman's darling?
Who wouldn't be a fireman's bride?
I'm going to be a fireman's lover,
I'm going to be a fireman's wife.

Probably the most colorful activity of fire companies is the parade. In 1824 the firemen of Manhattan and Brooklyn put on a gala parade in honor of the visit of Lafayette. The engines formed a line in City Hall Park and the three longest ladders were set in a tripod with torches at the top; then the last two engines threw water to extinguish the torches. A year later the firemen marched to celebrate the opening of the Erie Canal—a challenge for other cities to surpass.

For the centennial of Washington's birth the Philadelphia firemen staged their first parade. New Orleans firemen in full uniform held a solemn procession on June 26, 1845 for the funeral of General Jackson.

Philip Howe described a scene in New York, June 3, 1850:

We witnessed last night an exhibition of a novel and pleasing description; a procession by torchlight of the New York firemen in their picturesque scarlet woolen shirts, with their engines decorated tastefully and brilliantly with festoons and wreaths of flowers. The occasion was the reception of one of the companies of Philadelphia firemen, and it is to be hoped that the decorous, orderly example of our boys may not be lost upon the proportion of the fraternity who are in the practice of making night hideous in our city of brotherly love.

In March 1849 the William Penn Hose Company of Philadelphia invited the Reading Hose Company No. 1 to parade with them, an invitation which was reciprocated for Reading's first firemen's parade in July 4 of that year. Ever since 1853 Reading firemen have put on a Labor Day parade.

On October 10, 1850 there was a tournament in Chicago to honor firemen of all the lake cities: Kenosha, Racine, Milwaukee, Detroit, Buffalo, etc.—the most splendid parade ever seen in the Midwest. There were 14 engines, two hose carts, a ladder truck, and 900 firemen with four bands. A pumping contest narrowed down to Buffalo's crack Engine No. 9 and Milwaukee's Oregon No. 3, the winner.

In Washington, D.C. the custom on May Day was to strip hose carriages early in the morning, then to go to the woods for loads of flowers. After decorating the carriages each man carrying a bouquet would parade singing. One song dedicated to the Franklin Company was "God Bless the Noble Fire Boy" by C. W. Tayleure. It became popular throughout the nation through the medium of Kunkle's Minstrels.

However, because Franklin Pierce failed to attend a party or even respond to the invitation, Washington firemen refused to march in his inaugural procession.

A New York parade of 1859 took several hours to pass a single point. Each company was preceded by a band and pulled its own engine. Probably the most spectacular parade of the century was that put on by 6,000 New York firemen, each carrying a torch, in honor of the visit in 1860 of the Prince of Wales who stood on a balcony to watch.*

Then as now firemen on gala occasions wore dress uniforms rather than turnout gear. As Currier and Ives lithographs show, the turnout gear by mid-century—hat, coat, and boots—had become very much like that worn today. But when in 1858 the Hibernia Company of Philadelphia went on an excursion that covered New York City; Brooklyn; Boston; Charlestown, Massachusetts; and Newark, New Jersey, they wore red shirts, white leather gloves, green hats, and capes. They participated in exhibitions, torchlight parades, fireworks, and sumptuous banquets.

---

*Possibly because of this experience he himself became a fireman in 1865 and kept a complete kit, helmet, boots, axe, etc. at the Chandos Street station, London. He actually fought fires.

Perhaps the most spectacular dress uniform was the Turkish style of the New York Zouaves—one they carried with them into the Union Army.

Not only the dress of nineteenth-century firemen but also their apparatus was designed for display. Engines and hose carriages were painted with scenes from mythology, literature, and history. The Franklin Company of Philadelphia had scenes from the life of Franklin: arrival in the city, sitting at his desk, serving as a fireman, flying his kite. Mythology gave painters an opportunity to depict feminine nudes. Well-known painters were employed: Thomas Sully, Henry Innman, John A. Woodside, Joseph H. Johnson, John Vanderlyn, and others. In 1839 the Neversink Company of Reading, Pennsylvania got a new engine decorated with a view of the city painted by Woodside. A company might pay as much as $1,000 for a painting on its engine. One company spent $500 for a single lamp; another had its brass work silver plated for $800.

Hats and speaking trumpets presented to fire chiefs were often expensive. The leading manufacturer of helmets, appropriately named Henry T. Gratacap,* made one with gold and silver mounting for Foreman Hall of Sacramento at a cost of $1,350.

Men paid for their own uniforms and raised money by fairs, assessments, and raffles to pay for the decoration of apparatus and presentation items. In parades they carried pieces of chamois to wipe off any dust from their machines.

All this display was not only for the home folks. As has been indicated, the phrase "visiting fireman" was no figure of speech. At various times District of Columbia companies

* He said of the firemen who hung out at his shop: "They would come there and blow fire-talk from daylight to ten o'clock at night. I gave them tobacco and water, but no liquor."

visited Baltimore, Philadelphia, and other cities. In return, companies from these cities and from as far away as Rochester came to Washington between 1849 and 1851. Reading and Philadelphia firemen visited back and forth, but Reading men went also to many other cities. In 1894 a large contingent went to Elmira, Geneva, Watkins Glen, Rochester, Buffalo, Niagara Falls, Altoona, Harrisburg, and Pittsburgh.

Fire company visits were usually reciprocated. As a rule the occasions were happy, but when the Veccacoe Engine Company of Philadelphia paid a visit to Washington, a fight broke out in the dining room of Willard's Hotel. Firearms were used and several men were injured, one shot through the breast.

A much happier hegira was that of two New Orleans companies to the Northern states in 1870. They visited Cincinnati, Baltimore, Philadelphia, and New York. So soon after the war their excursion took on to some degree the character of a diplomatic mission. As the New Orleans journalist George L. Catlin wrote from Cincinnati:

I have just witnessed a reunion more real, more practical, more productive in lasting results than ten years of Congressional legislation and tinkering at reconstruction laws could produce. I have just seen hearty Cincinnati firemen, who fought with Heinzelman, with Buell, with Hancock with a will clasp hands cordially, even enthusiastically with sturdy, robust firemen of the Southern metropolis, men who fought with a will with Lee, with Bruekner and with Hood . . . show me the American who could without a certain leaping of the heart see the whilom foes sitting down together to discuss stuffed crabs, Roederer and ice-cream. . . .

This visiting back and forth by no means stopped in the twentieth century. In 1923, to celebrate the 175th anniversary of the founding of the city, the Reading Hose Company entertained a company from Harrisburg with 70 men and a 30-piece band; one from Norristown with 123 men and a 30-piece band; and one from Mechanicsburg with 150 men and a 40-piece band—which must have almost depopulated that small borough. In addition to these units from Pennsylvania towns, 100 firemen came from Newburg, New York, and 50 from Portsmouth, Virginia. In 1925 the Neversink Company visited Cuba.

Today, with the universal use of motorized equipment, fire companies bring one or more pieces of apparatus from 50 or 75 miles away to attend a housing or anniversay celebration. In 1969 when the town of Bryn Mawr celebrated its centenary 40 companies paraded. The Gladwyne Company fed them ham, roast beef, corn on the cob, and 19,000 steamed clams. A parade in Wildwood, New Jersey in September 1976 had 1,500 pieces of apparatus and lasted from 9:30 A.M. to 6:30 P.M.

In New England and more recently elsewhere firemen's musters featuring old hand-pumped engines (of which more later) draw competing companies from several states.

Belonging to a volunteer fire company often became a family tradition almost like the inheritance of an acquired characteristic. Among the original 35 men newly appointed in New York in 1738 were two members of the Rome family. From that day on for 112 years there was no time when this name was not on the enrollments. In Reading, with a population of 2,500 in 1813, so many men from the Rainbow Company were in the war that the younger boys founded the Junior Company with the motto, "We strive to conquer and save." Many of them were sons of the Rainbow men. As of 1938 the fifth generation of one family was in the Rainbow ranks. Today it is still not unusual for sons of firemen to join the same company, or if they move away, to become a fireman in another community. The Gladwyne Company founded in 1944 has already had two generations of members from six families, and at present a seventh is represented by three generations of active firemen.

In the same tradition Beth Murine, a fire fighter in Eddystone, Pennsylvania, is the daughter of a retired fireman and sister of a lieutenant in the company.

Nor do old firemen necessarily fade away. In the days of hand-pumped engines the men had to be relatively young, but today many companies do not have a compulsory retirement age. In the suburbs older men are often available during the day when the younger firemen tend to be at work miles away. A few years ago a Reading fireman got a cut on the head, went to a hospital to have it dressed, then went back to fight the fire. The only remarkable thing about this was that the man was 89 years old. A picture taken at a Labor Day firemen's demonstration at Townsend, Massachusetts in 1950 shows an 89-year-old Ashland No. 1 engineer operating a 79-year-old steamer.

From early times there has been the problem of payments for injuries to firemen and death benefits to their families. Often this has been handled by means of voluntary contributions or some sort of benefit event. Even today the death of men during a fire, whether paid or volunteers, produces a fund-raising drive for the widows and orphans. But it has long been recognized that some more certain form of compensation is necessary.

Thus, in 1791 New York provided that the fines for chimney fires should go into a fund for disabled firemen. Two years later an amendment extended the relief to the families of firemen. In the 1830s the Philadelphia Association for the Relief of Disabled Firemen was organized for the purpose of "relief of disabled firemen, their widows and orphans, and of persons not firemen, who may sustain personal injury by fire apparatus." An act of 1857 appropriated to the Association the receipts from a 2 percent tax on the receipts of outside insurance companies doing business in the county and city of Philadelphia.

In 1834 New Orleans firemen organized the Firemen's Relief Association which later became the Fire Department and for a third of a century administered the city fire service as an independent body. There was no parallel of this in other cities.

The Philadelphia tax on insurance companies was declared unconstitutional by the State Supreme Court, not noted for liberal views, in 1861. However, beginning in 1879 a series of acts of assembly imposed a levy on out-of-state fire insurance companies to be allocated to municipalities. Later townships were included and still later the funds were mandated to firemen's relief associations.

This plan has been widely adopted throughout the United States, the levy of 2 or 2½ percent not necessarily confined to out-of-state companies.

The relief associations ordinarily provide injury and death benefits for volunteers; paid fire fighters are usually covered by workmen's compensation legislation and public employees' pension plans.

When a volunteer dies, he is often given a fireman's funeral. In the nineteenth century the hose cart was used as a hearse; today, at least in New England, the fire truck is used. In New York about 1,830 firemen bought six lots in Greenwood Cemetery and were presented with two more. They put up a 23-foot marble shaft costing $2,500—all paid for by the volunteers. The Fireman's Charitable Association of New Orleans used the proceeds of a bequest to establish the Cyprus Grove Cemetery, dedicated in 1841. The Jefferson

Fire Company No. 22 of that city had a battery of tombs topped by a bas-relief of a hand-pumped engine.

Responses to a questionnaire in 1976 showed that many fire companies across the nation observe some sort of funeral ceremony with the members wearing dress uniforms. A deceased auto mechanic may receive a 2-inch obituary in the paper; a faculty member will be memorialized in a formal resolution; but if either has been a fireman he can expect full honors and possibly a plaque in the firehouse inscribed with his name.

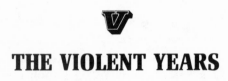

THE VIOLENT YEARS

As THE FOREGOING ACCOUNT shows, firemen from different cities came to each other's aid in big blazes, and fraternized at other times. However, during the middle decades of the nineteenth century volunteer companies in the same city engaged in a state of war. The era of the violent volunteers is notorious, but it must be considered in the context of the civic and national violence: anti-Negro and anti-Irish riots in Northern cities, the mayhem in the South against abolitionists, "bleeding Kansas," John Brown's raid, and a great Civil War. More on this later.

It is not to be wondered at that in an era marked by civic violence fire companies themselves became a disruptive element. In large cities, with the tendency of influential citizens to live in enclaves of handsome houses and for the proletariat to be huddled together, the membership of volunteer fire companies changed. The type of prominent citizens who had once belonged now let the lower classes do the dirty work. In smaller towns and cities the volunteer companies then as now might be manned by a cross section of the population, but it was no longer so in the big cities. A wealthy George Templeton Strong might run to fires but he did not pump an engine.

Furthermore, the proliferation of fire companies led to rivalries. A somewhat parallel situation today is that warring gangs are a big-city phenomenon; they do not happen in Podunk. A town with only one fire company did not, as a rule, experience warfare among firemen.

Trouble started early. In 1767 it is recorded that several Boston fire companies played their engines on each other. There are references to companies being disbanded for misconduct. After an 1837 clash between firemen and an Irish funeral procession, public sentiment turned against the firemen and Mayor Lyman took the first steps toward replacing the volunteer system with a paid department. The disbanded firemen established an anti-Catholic paper which during its brief existence attacked the Irish and the "paid Patriots" who replaced the volunteers. However, a volunteer element remained in the Boston Fire Department until 1873.

In 1850 Boston's chief engineer, Big Bill Barnicoat, was several times accused of intoxication at fires. City Council dismissed the charges, but almost every month at least one engine company was disbanded for fighting or most often for drunkenness.

To deal with similar disorders in Baltimore a delegation from all the fire companies met in 1834 to organize the Baltimore United Fire Department. Fire districts were laid out and assigned. Because rowdies could pose as firemen, it was resolved that every fireman be identified by a helmet.

Even so, thieves dressed as firemen went to fires to loot buildings. The firemen caught some of them and threw them into the Harbor Basin.

By 1847 disorder and riot had reached such an acute state that the Department met to devise means of stopping it. A resolution requested the mayor to lock the doors of engine houses whose members engaged in fighting. A bad fire the next year caused the mayor to unlock several engine houses, but many exiled firemen refused to work. An extensive area burned. Almost every Sunday there were disgraceful riots. It was usual for members of one company to set trash fires in a rival's district.

About midnight on October 8, 1858 there was a desperate struggle between the Rip Rap Club and the New Market Company. Pistol shots made it sound like a full-scale battle and there were a great many casualties, both killed and wounded. This led to the creation of a paid department March 30, 1859.

In New York in the 1830s, with 1,500 volunteer fire fighters, the East Side rowdies in the department were engaged in drunken brawls and street riots. When in 1831 James Gulick became chief, he tried to put an end to the rivalries and keep the toughs in line. Despite broad hints by the Tammany machine to lay off, Gulick went ahead. During a fire a messenger came with word that the aldermen had relieved him of his duties. He turned his hat around and started to leave. The firemen then stopped pumping and turned their hats around.

Mayor Lawrence rushed to the scene and ordered the men back to work. When they booed and jeered him, he drove back to city hall and demanded that the firemen be made to learn who was boss of the city. In the meantime, with the fire out of control, Gulick was persuaded to return. The men gave a mighty cheer and went back to work.

However, Gulick was definitely out. On the following day 800 firemen resigned in protest. For a year the department was badly undermanned, with only half the men needed to operate the equipment. Backed by the Resigned Firemen's Association, Gulick ran for city register and beat Tammany by the largest plurality up to that time. As will appear, volunteer firemen have been a potent political force from early times to the present.

Like New York, Philadelphia had a long history of fire company strife. In an era of scarce fire hydrants, fights tended to develop over the possession of one of them.* An ordinance of 1837 withheld funds from any company which tied up a second plug before hose had been led out from the first engine. Another ordinance of 1840 attempted to control violence and disorder by giving police power to a member from each company. A special committee recommended such things as limiting the membership of individual companies and prohibiting anyone under 21 to be a fireman.

In his history of the Philadelphia Fire Department J. Albert Cassedy, describing the late 1840s, wrote: "The spirit of disorder and misrule which had been growing annually for fifteen or sixteen years, was now at its height." The city's adjoining independent districts were a protection for gangs like Killers, Blood Tubs, Rats, and Schuylkill Rangers. On one occasion the carriage of the Franklin Hose Company was seized by a gang and thrown into the Delaware River. Moymensing Hose was then attacked by adherents of the Franklin.

The ballads at the end of this chapter give some of the

---

* In England rival insurance brigades often fought over a source of water while a fire burned unchecked. What made things worse was the insurance company practice of offering a reward to the first brigade at a fire. Thus a crew would attempt "to nick" the wheel of a rival engine.

Blackstone, p. 71.

flavor of the rivalries of the old Philadelphia companies and their hangers-on.

In 1853, at the fiftieth anniversary banquet of the Philadelphia Hose Company, Major Peter Fritz, who had been a member for 30 years, made a speech defending the firemen. Saying he had observed fire departments elsewhere he went on, "I therefore must confess that candor and truth oblige me to give preference to our own volunteer system." With certain reforms it would be "safer and better than any paid department that could well be devised."

In a burst of rhetoric he declaimed:

The quiet and unobtrusive citizen, who seldom goes to fires, can but imperfectly estimate or appreciate the perils, the hardships and services of a fireman's useful and glorious career. This peaceful citizen hears with horror of a disturbance or riot, that some beligerent fireman may, or may not be in some way implicated, but perhaps the fracas may have been caused by some rowdy hanger-on; and forsooth, this citizen stands ready to pull down a long tried volunteer system, and build upon its ruins another of doubtful expediency, because a paid department has been made to answer in other parts of our country. . . .

A well disciplined army cannot be held responsible for the sins and outrages of vile persons not belonging to its ranks; neither can a properly organized volunteer fire department be responsible for the iniquity, outrages and crimes that bad citizens may choose to commit.

He charged that much of the trouble was due to the inefficiency and negligence of part of the police force. Put in intelligent, sober, and honest policemen and one would hear no more of firemen's riots.

Major Fritz had a point in blaming part of the trouble on

rowdy hangers-on. Just as the campus riots of the 1960s were aided and abetted by the "street people" who hung around universities, the fights involving fire companies were at least in part caused by rowdy adherents of various companies.

The consolidation of the city districts in 1859 solved some of the problem by centralizing police and fire authority, but the agitation for a paid department continued. In 1868, when a bill to create one was before the Select Council, a crowd of firemen and friends filled the gallery and threatened the proponents. William Stokley, later Director of Public Safety, delivered a bitter speech. Pointing to the gallery he recalled that a volunteer had started an oil fire, and charged that some men in volunteer companies were professional incendiaries for the purpose of plunder and theft. After a storm of protest he went on: "Ruffians, that you are, you have unconsciously done the city a great good. By your very actions this afternoon you have proved the truth of every charge that I have made."

As Cassedy noted, "The volunteer department was a power." It was no child's play to destroy an organization which the habits and need of years had made a living thing, and which was endeared to the people by acts of noblest heroism. A paid department in Philadelphia did not come until 1871.

Before that a number of other cities had them: Cincinnati in 1852; Providence, Rhode Island in 1854; New Orleans in 1855 (disbanded that same year); Chicago in 1858; and New York in 1865.

Some, but not all, of these paid departments were at least partly the result of disorders by volunteers. Certainly cities other than New York, Philadelphia, and Baltimore experienced fire company strife. During the administration of President Tyler members of the Franklin and Anacostia

Companies got into a brawl in front of the White House. The armed conflict in Willard's Hotel between a Philadelphia and a Washington company has been recorded.

In Cincinnati fire company rioting became such a scandal that the citizens and authorities bought a steam pumper in 1852. It was designed and built by Moses, Alexander, and Finley Latta, the first builders of steamers in America. At its first appearance at a fire, it aroused the anger of the volunteers who cut hoses and hurled stones. Their hostility here and elsewhere was due to their well-grounded fear that they would be displaced. The citizens beat off the thugs, and the engine threw four streams on the blaze to the cheers of the crowd. So successful was it that Cincinnati became the first all-steam fire department in the world. These early Latta engines weighed five or ten tons, but could get up steam in four minutes and throw water nearly 300 feet.

With the advent of the steamer a large force of men was no longer necessary, thus making it possible for cities to introduce paid departments consisting of a relatively small number of men.

However, not all cities experienced fire company riots. Thomas O'Connor in his *History of the Fire Department of New Orleans* states that because the membership of volunteer companies there never passed into the hands of rowdies, the city was free from disgraceful riots. There were rivalries, as is shown by the picture of a gentlemen in a silk hat, sitting on a barrel covering a hydrant until the arrival of Company No. 5; but the historian insists that there were no "hateful feuds." (In New York a man who tried the barrel trick was beaten up.)

In his history of San Francisco Roger W. Lotchin states that the volunteer department there of the 1850s was usually free from the riots which plagued Eastern cities. It won high praise from the citizenry for its efficiency, morale, and courage. If Robert S. Lammont, a volunteer, is to be believed, the men after a fire, instead of stopping for a fight, would file past each other with such cries as "Hurrah for the Howard! She's always the first;" "Three cheers for the California—she is *some* at a fire;" "There comes the Monumental!" "Good for the Baltimorians. . . ."

The problem in that city was the maldistribution of services: most units and the cisterns were concentrated close to the business district, a situation which persisted into the 1860s. In the holocaust of 1851 all the cisterns were empty. By 1855 there were 50 cisterns, and a new chief tightly structured the volunteer fire department using New York as a model. Many individual companies, however, copied their regulations from Boston, Baltimore, and Philadelphia.

Politics entered the picture: the Democrats and Know-Nothings in the council deadlocked over appropriations to their favorite companies. Volunteers were in a continual struggle to increase their independence from politicians and the city government.

San Francisco's relative freedom from the violence that afflicted Eastern cities may have been largely due to its relatively small size: a population of 459 in 1847, 34,000 in 1852, and 56,000 in 1860.* The first engine company was organized in 1849 by David E. Broderick, who had been foreman of Engine Company No. 34 in New York. Using the fire company as a political base, he was elected to the state senate a year after he arrived, and in 1856 was elected to the United States Senate.

An anti-slavery man, he was challenged by a former Southerner, Judge David S. Terry, who killed Broderick in a pistol duel, The New York firemen held an imposing memorial service. A play about him, *Three Eras in the Life of a New York Fireman*, had a successful run in the Bowery Theatre.

---

*By 1835 New York's population was 250,000.

The Fireman, No. 1. Hand-colored lithograph published by Harrison & Weightman, Philadelphia, 1858. *(Courtesy of Mutual Assurance Company)*

Always Ready. Nathaniel Currier, himself a fireman, posed for this 1858 Currier & Ives print.

The Fireman No. 2. Hand-colored lithograph published by Harrison & Weightman, Philadelphia, 1858. *(Courtesy of Mutual Assurance Company)*

Facing the Enemy, from *Life of a Fireman*, Currier & Ives print.

Steam and Muscle from *Life of a Fireman*, Currier & Ives. *(Courtesy INA Corporation Museum)*

The Rescue, from *Life of a Fireman*, Currier & Ives.

The Fire Zouaves of New York saving Willard's Hotel in Washington, D.C., May, 1861.

Memorial to William Kelly, died 1862, aged 20. Cypress Grove Cemetery, New Orleans, founded by The Firemen's Charitable Association. *(Courtesy of Leonard V. Huber)*

Chicago, which was also relatively free from rioting by firemen, got its first effective volunteer company in 1835 at a time New York had 1,500 firemen. Eventually Chicago had 33 volunteer companies when, with the introduction of steamers in 1858, the city got a paid department.

In those cities which experienced violence, fire company bards produced battle songs. One in New York went:

The silver hook and ladder
  The pretty, golden Four,
To make Thirty-one the madder,
  Wash the paint from off her door.

And another:

Number Six has come on deck
With a new assistant sec.,
  Do ye mind?
He's as dirty as its water.

Tho' he thinks himself a snorter,
But he really hadn't orghter,
  Do ye mind?

These were mild compared to Philadelphia ballads such as:

OUTLAWS NO. 1

When first bright fate decreed our doom,
  And stamped the Outlaws on our hearts;
When joy and home did round as bloom,
  We swore they never should depart.

But serpent like the Bleeders came
To mar the social happy hour,
And rob us of our hard-won fame,
The Outlaws' Honor, Zeal and Power;

The Bleeders are a low-born race,
The scum and dregs of all the earth,
Who never could an Outlaw face
Nor give a decent action birth.

But skulking like a hidden foe,
With half a stolen brick in hand,
They tried to strike a deadly blow
Upon our noble Spartan band. . . .

Half a dozen stanzas later:

About the fight in Callowhill
Above Eighth Street in our city,
The Bleeders got a bitter pill
The subject of this noble ditty.

The Outlaws drove them to and fro,
Like straw upon the wind,
At every jump they got a blow,
But dared not look behind.

With broken heads let Bleeders gas
About fighting and their treasure;
We each have got a pretty lass,
And spend our days in pleasure.

So Outlaws to your trust prove true,
And then you're sure to nick 'em;
And should they offer fight to you,
Why that's the time to lick 'em.

The rival Bleeders also had a song:

> *Come all you jolly rowdies that delight to fires run,*
> *And I'll tell you of a crowd of boys, Called Bleeders*
> *No. 1.*
> *At Eighth and Callowhill they stand, as everybody*
> *knows;*
> *They run the Western Engine; they are down on*
> *Goodwill Hose.*

Chorus: *We are the Saucy Bleeders,*
> *I suppose you all do know*
> *We'll stand by the Western*
> *Or to Moyamensing go.* *

> *They were standing on the corner, in quietness and*
> *peace,*
> *When a band of ruffians ran us off, they called*
> *themselves police.*

> *In Callowhill near Schuylkill Sixth was the hardest*
> *battle ground.*
> *They began to fire their pistols, which made us look*
> *around.*
> *Although they fired them freely, they were of no*
> *avail,*
> *But there wasn't a shot the Bleeders fired but what*
> *did tell its tale.*

> *The lieutenant with his posse they thought they*
> *could us fight.*
> *By taking us from the engine house to the station*
> *house that night.*
> *In the morning we were all held to bail, from a boy*
> *up to a man,*
> *Before the Court and jury our trial for to stand.*

* A prison in the Moyamnsing section of Philadelphia; in use until the 1960s.

Here's a health to the Western and all who honor
    their name,
And powder and shot to their enemies—I do allude
    the same—

And here's good health to the Bleeders—I hope
    you'll drink their toast;
Likewise unto the Bleeders' gals, for they are the
    Bleeders' boast.

RAMS'SONG

Come all you jolly Rams
    And listen to my song—
I'll tell you of a cowardly crowd,
    Called Scroungers, No. 1.

It was on the 20th of July,
    It was on a Thursday night,
The Scroungers came to the corner,
    The purpose to raise a fight.

They came around with mud and lime,
    Our name for to deface,
But we caught them just in time,
    And whipped them face to face.

They thought by laying back with bricks,
    They would make us run away,
But we found out their dirty tricks,
    And drove them without delay.

The Scroungers are a cowardly set,
    And truth to you I'll tell,
We never have fought them yet
    But what we gave them h-ll.

ANCIENT RAMS, NO. 2

We are the Ancient Rams,
*   Who never fear our foes,*
At the corner of Second and Wharton we stand
*   And run with the Wecca Hose.*

Then arouse ye gallant Rams,
*   And by the Wecca stand,*
And show unto our friends and foes
*   That we're a sporting band.*

Our foes are called the Scroungers,
*   A name we never fear,*
For when they see us ancient boys
*   They soon will disappear.*

On the first of September,
*   Upon a Wednesday night,*
They stood at Wharton and Rye Streets
*   To show the Rams fight.*

Here's death unto the Scroungers,
*   And all who love the name,*
That name will send them down below,
*   From where the Scroungers came.*

TORMENTORS NO. 2

We are the saucy Tormentors,
*   I suppose you all do know;*
We stand at St. John and Poplar Streets
*   And run the Indy' Hose.*

Chorus: *Our name it is Tormentors,*
                  We fear no cowardly foe;
                  We're bound to fight and have our rights,
                  *Or to Cherry Hill * we'll go.*

          We are a saucy party,
              Our Machine is very light,
          And if we catch the South Penn Hose
              We're sure to have a fight. —Chorus

          Remember Fourth and Wood streets
              Which everybody knows,
          The Bed Bugs were run into
              By the Independent Hose. —Chorus

          The Carrolls just got in service,
              As everybody knows,
          But if they don't keep a lookout
              She'll be taken by the Indy' Hose. —Chorus

WORMS NO. 1

          Come all you jolly firemen
          That run to fires now and then;
          Come all you Worms, face your foes,
          Especially the cowardly Warren Hose.

Chorus: *Sing tally-i-n, sing tally-i-a,*
                  The Philly to a friend will never say nay.

          You have heard of this company before,
          And the many things happened within the doors,
          Their patent leather shoes and linen coats,
          And hair upon their lips like Big Nanny Goats.

* The Eastern Penitentiary.

But when fires do break out,
You'll see the Philly bold and stout,
The men with the engine, the boys with the crab,*
I'll tell you it gets the Warren boys mad.

And if they see the crab so blue,
Follow it they will not do,
Nor will they start their carriage with fear,
Till they know the crab is nowhere near.

So then they stopped to raise a fuss,
But we were right in for a muss,
And there being no police about,
We routed and kicked them all about.

At the fire out Market street,
We took their plug from them so sweet,
We took off their hose and put on our'n,
And the way we waxed them made them frown.

Now Warren boys, you "Almshouse" trash,
If you don't look out that carriage we'll mash,
But with all your knives, your pistols and dirks,
We'll use our Fist to do our work.

---

* *crab:* a heavy type of four-wheeled hose cart.

VI

ARSON, RIOT, AND WAR

THE BATTLES BETWEEN RIVAL fire companies were a serious disturbance of the peace, but the riots, usually against blacks or the Irish, were murderous. Firemen called in during the frequent incidents of arson were occasional victims.

One of the first riots was in Philadelphia in August 1824 when Southern sympathizers and blacks got into a fight at an exhibition of flying horses (merry-go-rounds) attended by people of both races. The building and its machinery were destroyed. Next morning a crowd went to the Negro section, broke down doors, and destroyed furniture. Another mob tore down a Negro meeting house. Blacks who were caught were beaten. The police were unable to cope until 300 constables were sworn in. Mayor Swift managed to disperse a crowd which was about to tear down a Negro school.

Political rivalry could also produce disorder. On election

87

night October 14, 1832 the Whigs in the Moyamensing section drove the Jackson men from their headquarters tent. Reinforced by men from Northern Liberties, the Jackson men attacked the Whig headquarters where they were met by buckshot from the windows. The Jacksonians broke in and set a fire which spread to neighboring buildings and could be seen for miles. Like demonstrators of the 1960s, the mob attacked firemen and cut hoses, but the firemen kept at work in a vain effort to save a row of houses.

Violence spread to other cities. Between 1830 and 1850 there were 35 major riots in New York, Philadelphia, Boston, and Baltimore. Both the blacks and the Irish suffered. In 1834, following an attack by a black servant on a former U.S. consul in Trinidad, a Philadelphia mob attacked the Negro district and set fire to a row of houses. As usual the mob attempted to cut hose lines. The firemen fought off the attackers and saved all but one house. The next day hundreds of families moved out and camped in the woods and fields.

In 1836 a mob broke up an anti-slavery convention in Granville, Ohio; later, abolitionist editor Elijah P. Lovejoy was killed in Clinton, Illinois.

In Philadelphia one of the worst outrages was the burning of Pennsylvania Hall in 1838. It had been built the year before by abolitionists, largely Quakers, at a cost of $100,-000. On the first floor were offices and committee rooms; on the second an auditorium seating 3,000. Whites and blacks were seen walking together to meetings, and rumor had it that men and women of the different races walked arm in arm.

At a meeting on May 16, 1838 the speakers were William Lloyd Garrison, Maria Chapman, Abby Kelly, and Angelina Grimké. Stones were thrown through the windows but the speakers went ahead. On the 17th the manager decided to close the hall and give Mayor Swift the key. The mayor came with police, some of whom were knocked down. The crowd then set the place on fire and broke gas pipes inside it. Firemen were deterred by threats from the crowd. The owners sued the city and a jury of inquiry set the loss at only $33,000.

The next evening the Shelter for Colored Orphans run by the Society of Friends was attacked and set on fire. Again firemen were threatened but police magistrate Morton Mc-Michael called on citizens to aid, and under his leadership the Good Will Fire Company cleared out the rioters and saved the building.

Anti-Negro feeling was particularly strong in Phila-delphia, possibly encouraged by a merchant class which did much business in the South. Also, it is probable that the city had the highest proportion of blacks anywhere in the North. There were anti-Negro riots again in 1840, 1842, and 1848. As usual these involved arson and attacks on firemen, at least one of whom was killed. It is worth noting that during the mob violence against blacks, the volunteer firemen of Philadelphia were on the side of law and order.

This was not always the case with firemen during the riots involving the Irish. The massive immigration, especially during the famine years 1845-1848, caused ill feeling from Baltimore to Boston, and led to the formation of the Native American Party (Know-Nothings) with its center of activity in Philadelphia. The party soon had branches in Baltimore and New York. It held its first convention in 1845 in Har-risburg, Pennsylvania.

Before that there had been the burning of the Ursuline Convent in Charlestown, Massachusetts in 1837. It had been provoked by Rebecca Reed's lurid stories of convent life and by bickering over a Catholic cemetery at Bunker Hill. The refusal of the mother superior to permit the Charlestown

Selectmen to investigate the purported existence of dungeons and torture chambers inflamed Charlestown truckmen and New Hampshire Scotch-Irish bricklayers who led a mob against the place. The mother superior unwisely threatened to ask the bishop to send 20,000 Irish defenders.

The press condemned the burning of the building and a mass meeting in Faneuil Hall expressed sympathy for the victims, and resolved "to unite with our Catholic brethren in protecting their persons and property and civil and religious rights." However, the Montgomery Guards, an Irish military company, was attacked by the men of the Boston City Guards, who refused to parade with them.

In New York, Samuel F. B. Morse wrote anti-Irish articles for his brother's newspaper. The ill feeling generated by the Know-Nothings led East Side toughs, mostly Irish, to lie in wait to pounce on firemen on their way to a blaze.

An especially violent outbreak occurred in the Kensington section of Philadelphia on May 4, 1844 when Irishmen in the Hibernia Hose Company (not to be confused with the older and largely Protestant Hibernia Fire Company) fired into a meeting of Native Americans. In the ensuing riots several persons were killed and over 50 wounded. The Hibernia Hose house was stripped and the apparatus demolished. The sheriff called for the military,' but the officers refused to respond. However, on Tuesday the First City Troop, an old and aristocratic militia company, did turn out.

On the following day a mass meeting was fired upon from the ruins of the Hibernia Hose house. The mob set the ruins on fire, a blaze which spread to a row of houses, destroying about 30. At 2 in the morning the mob set fire to St. Michael's Church with the help of the colonel assigned to protect it. At the same time the Female Seminary at Second and Phoenix Streets was burned. Then the Church of St. Augustine, the largest Catholic church in the city, was set on fire. Firemen made no effort to save it but concentrated their efforts on adjoining property. Even so, a large building used as a schoolhouse and containing a large, costly theological library was destroyed along with rare books.

Peace was restored by deputized citizens and the military, but a grand jury blamed the disturbances on "the efforts of a portion of the community to exclude the Bible from the public schools." It charged that a party in the peaceful exercise of the sacred rights and privileges guaranteed to every citizen by the Constitution "had been rudely disturbed and fired upon by a band of lawless, irresponsible men, some of whom had resided in the country for only a short time." In any case the incident led to a huge increase in membership in the Native American Party.

One problem was that before the consolidation of the city the police in various districts could not cross boundary lines.

A different situation existed in Milwaukee where before 1854 the volunteer firemen were the chief civil authority. Once they had caught a pair of murderers. In 1851 firemen were called out to quell that so-called Leahy Riot. Leahy, a lecturer billing himself as "the Reformed Monk of La Trappe," proposed to reveal sordid facts of monastery life. A crowd of Irish Catholics gathered to prevent the talk, which was to be given in a Methodist church. Firemen summoned by the mayor formed a hollow square and led Leahy to a hotel. The good nature of the crowd was aroused by men armed "with faucets instead of pistols."

To an astounding degree the old-time firemen had to cope with arson. In addition to incendiarism growing out of riots there were many private acts. In 1654 Boston established the death penalty for arson; nevertheless, some 25 years later certain persons were under vehement suspicion of attempting to burn the town. They succeeded in burning one house.

The fire of September 21, 1776 which destroyed 493 houses in New York—one fourth of the town—was believed to have been set, either by the British or by patriots who wanted to thwart the occupation. Because of wartime conditions the engines were out of order and there were few hands to fight the fire. During 1803 there were two attempts to burn the city of Philadelphia. In addition there were a large number of incendiary fires, one causing the death of three men under a falling wall.

In 1852 there were three separate attempts to burn San Francisco, and as late as 1856-1860 arson still made up 25 percent of the causes of fire there. A Baltimore incendiary in 1857 caused a fire by throwing a pail of blazing coal oil into a store—a conflagration that involved all the fire companies in the city.

The 1861 account of New York firemen published in *All the Year Round* and quoted previously added the statement that "The special disgrace of the city is the incessant occurrence of incendiary fires."

On February 10, 1864 a discharged coachman set fire to the private stable of Abraham Lincoln, killing six horses.

The most serious fire in Philadelphia up to February 8, 1865 was set by a volunteer fireman for the purpose of having a fight with a rival company. It involved a coal oil establishment with 1,500 barrels of petroleum which poured down the street, causing 50 other buildings to burn. Six people burned to death and 100 were left homeless.

Arson is still a major cause of fires. The figures for Oregon, which publishes detailed fire statistics, showed that in 1974 incendiarism was the cause of 10.5 percent of all fires, second only to 12.19 percent due to careless smoking. The Michigan Fire Fighters Training Council in 1975 estimated that more than 30 percent of all building fires were now being set. It spoke of "The fantastic increase in arson. . . ."

Modern incendiary fires are often set to collect insurance,

and tend to increase during depressions, but as in the past some are set for revenge or for a thrill.

As is suggested by the New York fire of 1776, wartime conditions give rise to arson, and many firemen were likely to be in the army. During the first year of the American Revolution the New York firemen wrote to the Provincial Congress: "We are willing to serve as firemen; and if a general attack should be made upon the city, we are willing and ready . . . to turn out as soldiers."

Early in the Revolution, Lambert Cadwallader raised a Philadephia company of soldiers from Engine Company No. 87, known as the silk-stocking company because of its aristocratic membership. In Reading, with a population of a little over 500, Capt. Joseph Hiester at his own expense organized a military company made up of members of the Rainbow Fire Company.

In the War of 1812 so many Rainbow men were in the army that the younger boys founded the Junior Company with the motto, "We strive to conquer and save."

Although the motto was in English it is probable that commands were given in Pennsylvania Dutch. During the Civil War, the Heidelberg Brigade, opposed to its continuance, armed themselves with spikes, scythes, pitchforks, and hatchets, and marched up Penn Street to the courthouse. When the police were unable to handle them, the Junior Fire Company hitched up a fire hose. The Heidelberg leader shouted, "Do cumpt de cannon; spring bower, spring!" (Here comes the cannon; jump boys, jump!) The Juniors scattered the invaders. Men from the Keystone Company guarded the bridge leading toward Harrisburg, and the Liberty Company held the Lancaster bridge. Some of the Heidelberg men were forced to swim the Schuylkill.

The most famous of the Civil War fire companies was the New York Zouaves. Ten days after Lincoln's call for volunteers New York firemen organized the First Fire Zouaves,

1,100 strong. Dressed in red shirts, gray jackets, and gray flowing trousers, they marched down Broadway to embark, cheered by a crowd of 20,000.

While they were bivouacked in the Capitol, in Washington, a building next to Willard's Hotel caught fire. Their commander, Col. Elmer E. Ellsworth, ordered 100 men to the blaze, but the whole regiment leaped out of the Capitol windows to fight the fire. An artist's sketch of the action shows men on the roof dangling a fireman by his feet while he squirts a hose into a window. A reporter wrote that "They destroyed nothing unnecessarily and nothing was missing."

Marching through Alexandria, Virginia they spotted a confederate flag on the Marshall House. Colonel Ellsworth dashed up the stairs, and was shot by the proprietor of the hotel. His body was taken to the White House to lie in state.

At the first Battle of Bull Run the Zouaves lost 200 men, killed, wounded, or missing. On September 13, 1861 the Secretary of War mustered them out.

However, a second body of Zouaves was recruited and fought at Antietam, Gettysburg, Fair Oaks, etc. Originally numbering 814 men they were down to 324 at Gettysburg, where they lost 162 more. A statue there shows a soldier and a fireman side by side.

When New Orleans fell into Federal hands in 1862, General Butler published an order saying: "The various companies comprising the Fire Department in New Orleans will be permitted to retain their organizations, and are to report to the office of the Provost Marshal, so that they may be known, and not interfered with in their duties." However, the firemen were unwilling to take the oath of allegiance required of all citizens. President Marks of the Fireman's Charitable Organization, which controlled the department, went to Butler with the proposal that the firemen be permitted to remain neutral. Butler promptly approved, and the firemen were permitted to draw up their own parole agreement.

When a cholera epidemic followed the war, the firemen used their engines to clean out the drains in the city. The council passed a resolution of thanks.

At the great Sanitary Fair in 1864 the New York Fire Department had one of the greatest attractions. An observer reported: "Nothing can exceed its magnificance. When lit up at night it is perfectly dazzling." It was a stand 60 feet long covered by a canopy decorated with hooks, ladders, lanterns, and hose, and with the motto "UNION" constructed of gas pipes. Shields bore the names of battles in which the Zouaves had fought. Eyewitnesses were also impressed by the "sprightly young ladies" in attendance.

To protect government property when Washington was threatened by the *Merrimac,* the Hibernia Company of Philadelphia came in 1862 with a fully equipped engine. Along with two other volunteer companies it was put under the command of General Meigs, the quartermaster. However, fights developed between the volunteer companies and the Government Department over which should be in charge at fires. The volunteers usually won. The trouble was finally settled by an agreement whereby the Government Department was in charge on government property and the volunteers at house fires.

During the three days of anti-draft rioting in New York, July 1863, incendiaries burned more than 50 buildings, including the Colored Orphan Asylum. The police were unable to control the mob, which hanged Negro men, women, and children from lamp posts. Firemen not only fought the blazes, but many members of the Exempt Engine Company stayed several days and nights in the Tribune building, to be ready at a moment's notice.

An attempt in 1864 by Southern sympathizers to burn

New York by setting a series of fires was thwarted by Chief Decker and the volunteer companies, which kept men on the watch.

It is obvious from all this that the volunteer firemen, despite their record of inter-company strife, performed much civic and national service. Their record of heroism in war and peace gained them much public support despite their often unseemly behavior.

Silver service presented to Newton Marks, president of the New Orleans Firemen's Charitable Association, by the firemen of that city in 1872. The whole set weighs 800 ounces of sterling silver. It is decorated with symbols of the craft: pipes, hoses, hydrants, axes, ropes, torches, helmets, hat fronts, trumpets, and hooks. *(Courtesy Firefighting Museum of the Home Insurance Company)*

Holding the hydrant. John Malloy of New Orleans Fire Company No. 5, keeping the hydrant covered with a barrel till his Company arrives. (History of the Fire Department of New Orleans, *Thomas O'Connor, 1895*)

1884 Button steamer used by the Hershey Fire Company in the early 1900s. *(Courtesy Hershey Museum of American Life)*

Early suction pumper.

Steamer with Christie tractor.

First self-propelled gasoline pumper, 1906, Radnor Fire Company. It had two engines.*(Courtesy Waterous Co.)*

Self-propelled one-engine pumper, 1907 *(Courtesy Waterous Co.)*

1926 American LaFrance Pumper used by Cape May Point, New Jersey volunteer fire company

A 1931 Ahrens-Fox 1000-gallon pumper, used in North Tarrytown, New York *(Courtesy Hall of Flame, Phoenix, Arizona)*

1941 Autocar pumper used by the Merion Fire Company, Ardmore, Pennsylvania *(Courtesy W. Robert Swartz, Ardmore)*

Housing of new pumper, Gladwyne, Pennsylvania, 1956

Autocar Fire, July 31, 1956, Ardmore, Pennsylvania.

Autocar Fire, July 31, 1956, Ardmore, Pennsylvania

Thomas Murray and Ernest Earnest douse the last of a housefire, Gladwyne, Pennsylvania, 1957.

VII

STEAM AND MUSCLE

FROM THE BEGINNING TO THE present, volunteer fire companies have been concerned with their apparatus. The pattern started early. As Chapter II shows, the fire engines bought by Boston, New York, and Philadelphia before the formation of volunteer companies almost invariably lapsed into a state of disrepair within a short time. On the other hand, the minutes of the early companies are sprinkled with notations of the appointment of committees "to enquire into the state of the engine," and with authorization for expenditures for repairs. Companies regularly fined members for failure to keep their buckets and salvage bags in repair.

As has been shown, the first engines, hose, and many of the buckets were imported, but American makers soon tried to improve on foreign models. Apparently the first engine built in Philadelphia was by Anthony Nichols who in 1735

103

asked for payment of £89.11.8. A committee appointed by the Council reported that it was very heavy, and probably would not last long. A more successful builder, Philip Mason, opened shop in 1768. By 1800 Philadelphia had four builders of fire apparatus and that city remained preeminent in the manufacture of engines for another 75 years.

The Union Fire Company, which had probably started by using city-owned engines, imported one in 1743. Less than two years later Benjamin Franklin paid £13.4 "for an Improvement to the Company's Fire Engine." The next month it was decided to buy six ladders. As early as 1761 the Hibernia Fire Company voted to pay a man for taking care of the engine, and seven years later to pay another man for keeping it in repair.

At the December 5, 1791 meeting of the Diligent Fire Company it was moved that they accept the proposal of Mr. Mason [?] for making an engine which would pump 170 gallons a minute and throw water 70 feet in a straight line *and shall far exceed any Engine ever in America.* [Ital. mine.] (That is not bad; some of the small engines today pump only 250 gallons a minute.) In 1794 Patrick Lyon invented an improved engine which would throw more water than any built hitherto. The Pennsylvania Company in 1807 paid $750 for a Lyon engine which could throw water 170 feet without spray. It was asserted to be the equal of any engine in Philadelphia. Lyon also made the first hose carriage and continued to build engines until 1854.

Late in 1790 the Delaware Fire Company postponed a discussion of a committee report on a "machine for conveying People and Goods from the upper Stories of Houses on Fire." It was finally rejected at the February meeting—probably because of its cost. This was almost certainly the "Speedy Elevator," an ancester of the snorkel * or "Cherry

---

* *snorkel*, a trade name for one brand of articulated boom, has become a kind of generic term like *thermos* for any vacuum bottle.

Picker." Its inventor, Nicholas Colin, D.D., read a description of it and presented a model of it to the American Philosophical Society on December 2, 1791, and was honored with the Magellan gold medal in 1795.

It consisted of a wheeled platform with two tall uprights between which there was a mast with an arm holding a basket. By means of a windlass the mast and basket could be raised and lowered. Although the Delaware Company rejected it, others apparently adopted the device.

Eight hundred feet of riveted hose, far superior to sewn hose, was obtained by the Resolution Company in 1811, the year it was invented by two members, Abraham L. Pennock and James Sellers. (The early companies had difficulty with the spelling, using variously, *hose, hoose, hooze,* and *hoase.)* Pennock and Sellers also invented a furnace for drying hose. This consisted of a brick tower filled with charcoal, above which was a wooden steeple in which the hose was hung to be dried in gradual heat.

Leather hose was always a problem. After a fire it not only had to be dried but also treated with neats-foot oil. In 1807 the Neptune Hose Company got a carriage with two rollers at the end to squeeze water from the hose. Five years later Perseverance Hose got a carriage on springs, then a second with a reel instead of a box bed.* About 1818 the Phoenix Company got an arched carriage for the reel—a design which lasted nearly a century.

The powerful Philadelphia-style engine with two banks of brakes fore and aft and weighing 4,000 to 4,800 pounds was widely used in Boston, New York, and other cities, but because it required 30 or more men to pull and pump it, not to speak of relief crews, it was not adapted for use by small departments. Its advantage was that it could be pumped in narrow city streets or alleys. Until the twentieth century

---

* After a century of reel-type hose carriers fire companies have gone back to the flat bed type, except for the small diameter booster line fed from a tank.

small places continued to use old-fashioned tub engines.

For one thing there was the problem of water supply. Cities installed cisterns, San Francisco as late as the 1850s. Early water pipes in Boston, New York, and Philadelphia were of wood and the first hydrants had no hose connection. Water was run into a pool or buckets and dipped or suctioned out. In Philadelphia there was a suggestion in 1803 that hose might be attached to hydrants and a standard size adopted.*

That city, the first in the country to have piped water, set up a waterworks in Center Square in 1801, but it did not furnish enough pressure for fire fighting. That came in 1815 with the building of the Fairmount Waterworks on the Schuylkill River and the introduction of cast iron pipe. The firemen no longer needed to use buckets. Reading, Pennsylvania had piped water in 1821 and a few fire hydrants.

As the account of New York's great fire of 1835 shows, that city's firemen had to depend on water pumped from the rivers. Therefore they celebrated the opening of the Croton Reservoir in 1842 with a gala parade. Chicago got its first piped water that same year.

Pittsburgh, which got its first volunteer company in 1793, had so many by 1845 that they could not all be supplied with water. In that year a combination of low water in the reservoir and a high wind led to a conflagration which between noon and 7 P.M. burned the best part of the city—some 982 blocks of buildings.

Before the introduction of piped water, or even later when there was an inadequate number of hydrants, not all engines could be brought to bear on a fire: they had to be hooked in relays—a source of conflict when one crew tried to wash the engine of another company. However, when one company

---

* Philadelphia and its suburban companies still use different-sized hydrant connections.

called "Will you take our water?" it was considered a disgrace if the other refused. In an era of scarce hydrants, fights over the possession of one were common.

The introduction of the hand-pumped engine had created the first great revolution in fire fighting. An engine company required more discipline and teamwork than a motley crowd with buckets. Engines required firehouses and engineers with mechanical skill to keep them in shape. As has been shown, pride in the engine or hose cart induced the cohesiveness and company spirit which sometimes led to excesses. Furthermore, the men who pulled and pumped an engine had to be young and muscular, that segment of the population which in all ages has been most ready for a fight in peace or war. In short, the development of the hand-pumped engine was at least partially responsible for the violent years.

Thus, the introduction of the steam pumper created a second revolution both technological and sociological. This helps to explain why volunteer firemen so long hospitable to or even enthusiastic about new or improved apparatus became early opponents of the steam fire engine. They quickly realized that it would make the large and brawny fire company obsolete. Instead of a crew of 30 with as many in relief to pump the engine, a steamer needed only an engineer and a man to fire the boiler. In a pinch one man could do both jobs and he need not be young—witness the 89-year-old operator mentioned on page 65.

Two characteristics of the steamer helped to force the employment of at least some paid men: the machines were usually too heavy to be pulled by hand, and their engineers had to be experts. Thus, although borrowed horses were sometimes used, most companies with a steamer put in a stable. This meant, of course, men on duty round-the-clock to feed and care for the animals. To keep a steamer from freezing up, a man had to maintain a heater or furnace in the firehouse. In big cities the furnace was hooked into the engine boiler with a snap connection to enable it to raise steam in a hurry.

The first steam fire engine was probably the "Novelty" built in London by George Braithwaite assisted by John Ericsson (of later *Monitor* fame) in 1829. In New York Paul Rapsey Hodge produced the first American steamer in 1841. Commissioned by the insurance companies, it was capable of throwing a stream to a height of 166 feet. However, it weighed over seven tons and looked rather like a locomotive because it was self-propelled in the same fashion.

It was assigned to Hose Company No. 28, whose members promptly objected. The firemen had some valid objections: the engine took a half hour to get up steam and it frequently broke down. Partly for these reasons and partly because the insurance companies feared reprisals by the volunteer firemen who tended to sabotage the early steamers, it was sold.

The first successful steamer in America was built by Moses Latta of Cincinnati in 1852. It could raise steam in 4 minutes, 10 seconds, and using 350 feet of hose, throw a stream 130 feet. (Friction loss in hose causes a drop of about 5 pounds of pressure per 100 feet.) As has been noted, when it was used at a warehouse fire, the volunteers tried to cut the hose lines, but were prevented by the crowd. This led directly to the purchase of more steamers and the institution of a paid department.

In 1855 another Latta steamer, the "Miles Greenwood," was demonstrated in a contest with hand engines which threw water a longer distance. Rejected in Philadelphia and New York, it was bought by Boston, which found it too large for the narrow streets.

Because of opposition by firemen both New York and Philadelphia were behind Cincinnati in converting to steam. A group of Philadelphia citizens raised money to present the city with a Latta engine in 1855. As Cassedy commented: The city "found itself in the position of a man who has won an elephant in a raffle." It weighed 20,000 pounds; there were no horses to pull it; an engineer and assistant had to be paid salaries; and in three years it cost $20,000 in maintenance and repairs.

However, the tide was changing. In 1854 the Common Council of New York City asked Chief Corson for a report on the expediency of steamers. He cautiously replied that he "entertained favorable views as to the propriety of introducing steam power. . . ." and invited the Council to a demonstration of one—possibly the "Miles Greenwood"—on its way to another city.* He advocated the appointment of four paid men to take charge of each engine: two to handle the horses and two to be in charge of the engines—in reality a miniature paid department.

In an 1856 contest between a Latta engine and the "Hay-Wagon," a big hand pumper, the latter won, but observers noted that at the end of the contest the Hay-Wagon men were exhausted, whereas the steamer was still ready to work. When the Common Council bought two steamers for $17,000 each and put them in service in 1859, one fireman remarked that the volunteers "began to realize what fools they had been for not introducing steam themselves."

On May 18, 1858 Capt. John Ericsson docked his steamboat in Baltimore and mentioned that he had on board a steamer bound for Norfolk. The crowd demanded a demonstration. Approving the machine, the members of the First

---

* In London, despite the fact that Braithwaite's engine had worked continuously for 5 hours at a fire February 1830, while hand engines froze up, the British did not accept steamers until 1858. James Braidwood, who in 1833 had been appointed chief officer of the amalgamated London insurance brigades, opposed steamers for 20 years as did other chiefs. Opposition also came from the retained (U.S. "on-call") men who received free beer and a few shillings for pumping. They often worked the brakes to the chant of "Beer oh, Beer oh," and if the supply were slow in coming would simply quit work. Anxious to preserve their pay and their beer, the mob would sometimes cut the hoses of steamers. Blackstone, pp. 114-15.

Baltimore Hose Company bought it on the spot. Then because of its tall stack they couldn't get it into the firehouse. After guarding it all night they got a mechanic to put the stack on a hinge.*

At a banquet in New York in November of that year, held in honor of the visiting Hibernia Fire Company of Philadelphia, a toast was offered to "The Steam Fire-engine—the greatest auxiliary to a Volunteer Department."

A year earlier the Philadelphia Hose Company had resolved to have a steamer of its own. The Hope Company got one in 1858 and Hibernia in 1859, by which time there were 20 steamers in the city. As Philadelphia did not get a paid department until 1870, this means that the steamers were operated by volunteer companies supplemented by paid hostlers and engineers.

Steamers ran into opposition not only in New York and Philadelphia but also in New Orleans, and, as usual, the technical questions got mixed up in volunteer politics. A group of New Orleans underwriters brought in a Latta engine, "The Young America," in 1855 and the same year the city council voted to have a paid department. The volunteer firemen revolted and brought 28 hand engines to Lafayette Square where they surrendered them to the city. The "Young America" was one of the monster early Latta engines weighing 9 tons. Eventually the chief advised the council "that costly and expensive piece of machinery has, in my opinion, totally failed to perform anything like proportionate service to the money annually expended on it by the contractor, say six thousand dollars."

The new paid men were green and the old timers made fun of them with the result that the fire record became

---

* A common problem today is that modern apparatus is often too tall or wide for firehouses built 30 or 40 years ago. More than one company has had to put in larger bays.

appalling. City Council then provided that a contract for fire protection be let. The underwriters bid $100,000 but the Firemen's Charitable Association (FCA) made up of the volunteers bid $85,000. In a published address to the city they stated, "The volunteer firemen can furnish a force of *fifteen hundred men* and *twenty-three companies* for an annual appropriation of fifty thousand dollars." The underwriters then put in a lower bid, but the FCA came down to $70,000 and offered another $70,000 for the apparatus belonging to the city. They got the contract and ran the department for 36 years.

By 1861 six New Orleans companies had converted to steam. This was happening all over: Brooklyn, Rochester, and Chicago got steamers in 1858; Milwaukee in 1861. Unlike New Orleans, all these cities instituted paid departments soon after steamers were procured.

In 1866 Daniel Hayes, a salesman for the Amoskeag Manufacturing Company of Manchester, New Hampshire rounded the Horn in a clipper ship carrying an Amoskeag steamer to San Francisco. He stayed to instruct the firemen in its use and became the master mechanic in the department. By 1870 he had perfected the Hayes Aerial, the first aerial ladder truck in America. It became popular from coast to coast.

The early Amoskeag steamers were remarkably efficient and much lighter in weight than the Latta machines. The first Amoskeag engine, designed by N. S. Bean, was tested July 4, 1859. Seven minutes after the fire was lit it had two streams playing to a height of 203 feet. By 1861 the company produced a small model weighing only 4,000 pounds— less than some of the Philadelphia-style hand engines.

This meant that they could be used by volunteer companies because they could be kept ready with shavings and kindling wood in the firebox, and if necessary be pulled by hand. Self-propelled steamers were unwieldy and not widely used, although Boston kept some in operation until after 1900.

Although the advent of the steamer led to the establishment of paid departments in a number of cities, this was not universal. In a number of places the change was partial. By 1863 Rochester, New York had four steamers serviced by a small paid force supplemented by one volunteer ladder and two hose companies. This mixture of paid and volunteer firemen is still common in medium-size communities.

Politics, of course, entered the picture. Volunteer firemen with their strong sense of cohesiveness formed a powerful political force. The New Orleans experience is a case in point. In New York City, William Marcy Tweed joined with some politicians who were organizing a fire company. It was Tweed who suggested the name, Americus Engine Company No. 6, and who gave it the symbol of a snarling tiger—the symbol Thomas Nast attached to Tammany Hall. It was Tweed who, as foreman in 1849, became a dashing hero in red shirt and white firecoat, and made No. 6, soon called the "Big Six," into one of the most famous fire companies of the city.

Tweed used the fire company as his stepping-stone into politics. Defeated for alderman in 1850, he won the post the next year. And as he later said, "There never was a time you couldn't buy the Board of Aldermen." Chief Corson once expelled him for an attack on Hose Company No. 31 but Tammany restored him as foreman of Big Six. It was Tammany influence which kept New York from getting a paid department until 1865.

Three years later in Philadelphia when a bill providing for a paid department was before the Select Council, a crowd of firemen and their friends filled the gallery and threatened the proponents. William Stokley, later Director of Public

Safety, delivered the bitter speech cited previously. Despite violent opposition an ordinance was finally passed in 1870 creating a paid department which went into operation the following January.

Attempts in other cities to establish paid departments often failed. In 1914 the City Council of Reading, Pennsylvania passed an ordinance creating a paid department, but the action aroused so much protest that a referendum was held. The measure was defeated 9-1, demonstrating the strength of the Firemen's Union, which to this day establishes rules and regulations, controls membership, and determines the duties of paid drivers. An attempt to set up a paid department during the 1920s in Harrisburg, Pennsylvania was also defeated because of opposition by the volunteer firemen. In recent years, however, the migration to the suburbs has so depleted the volunteer contingent that the department is largely professional.

Volunteers often became expert at handling steamers, just as today they have mastered the complexities of modern pumpers and aerial ladders. At the Paris Exposition of 1889, the American La France Company while demonstrating a steamer had difficulty in firing it. Along came a Reading fireman who volunteered to do the job. The American La France Company was awarded the first medal of honor for steam fire apparatus. In appreciation the company gave the Neversink Company of Reading a low price on a machine of special construction.

At the International Fire Congress of 1900 in Paris a Kansas City team of 14 volunteers competing with 8,000 contestants from 20 countries won a silver trophy and 800 francs for throwing a 1¼-inch stream 180 feet vertically and 310 feet horizontally.

As a rule, small-town companies could not afford to buy or maintain steamers. Over the years they had bought secondhand tub engines from city fire departments. Illustrations in the *Firehouse History* of the Townsend, Massachusetts volunteer department show about 50 hand tubs used in contemporary musters or in museums and dating from 1780 to the 1880s, all of them of small or moderate size. The huge Philadelphia-style engine was not practical for smaller communities.

However, the invention in 1867 of the chemical engine gave small-town companies a useful new piece of equipment. It consisted of a light carriage carrying two tanks either already full of water or filled on the scene. Soda was dumped in one, acid in the other; the solutions mixed at the pump. This produced a pressurized stream as long as the ingredients lasted. They could be renewed or the engine pumped by hand.

Obviously the chemical engine was not adapted for fighting large fires but it was a useful piece for a first attack.

As relatively small towns began to install piped water systems in the latter part of the century, volunteer companies could often get enough pressure directly from a hydrant to attack a house fire. Thus a hose cart alone was an effective piece of equipment.

But as a later chapter will show, the development of motor-driven apparatus produced the third revolution in fire fighting—one which made possible the modern volunteer company.

VIII

THE TECHNOLOGICAL REVOLUTION

THE INTRODUCTION OF gasoline-powered fire engines produced the third great revolution in fire fighting. Just as the
steamer had tended to drive out the volunteer companies,
the development of motor-driven apparatus has brought
them back into full flower. It was a development that accompanied the explosive growth of the suburbs—also in large
part a creation of the motor car.

The advantages of motor-driven fire engines are obvious,
especially for volunteer companies. Instead of a wait for a
crew to pull and pump the engine, or in small communities
for someone to get hold of horses from a livery stable or an
undertaker, the first four or five men to answer an alarm
could crank up the machine and be on the way. Other firemen could get hold of an automobile to follow the engine. It
is the practice followed today, except that even small com-

115

panies now tend to own several pieces of apparatus. Thus later arrivals can now man a second or third truck.

The transition was not made all at once. The earliest motor-driven type of apparatus, such as the American La France in 1903 and the Pope Hartford in 1909, was a chemical engine and hose wagon. City departments with steamers and long-ladder trucks often substituted a motor tractor such as the Christie (a trade name) for horses. For one thing, even after gasoline-powered pumpers were introduced about 1906,* they were less powerful than the old steamers. In 1892 the fire commissioner of Detroit wrote an enthusiastic letter to the Manchester Locomotive Works stating that in a test an Amoskeag steamer at top speed had pumped 1,875 gallons in 50 seconds; at ordinary speed the same amount in 57 seconds.

That was more than the rated capacity of the machine, but in the 1890s both Amoskeag and American La France offered 1,100 gallon-a-minute steamers. On the other hand, a huge Seagrave gasoline pumper of 1919 was rated at only 750 gallons a minute. In the 1920s and '30s pumpers rated at 350 or 500 gallons per minute were usual except in big city departments, which had more powerful ones.

Obviously small towns did not immediately buy motorized equipment; that became common only in the 1920s. However, a chemical engine or a hose cart could be hitched to the back of a fireman's car or pickup truck, or the hose could be loaded on the back of a truck. In Hummelstown, Pennsylvania the engine was once hitched onto the back of a Hershey trolley car.

Still another development made it possible for companies in small communities to arrive in time to do some good: that

* The volunteer Radnor Fire Co. of Wayne, Pennsylvania had one made that year by the Waterous Engine Works of St. Paul, Minnesota.

was the growing use of the telephone. Before that, cities had adopted the fire alarm telegraph invented in 1851 by Dr. William F. Channing of Boston. As such a system was impractical in smaller places, it was only after the spread of the telephone that an alarm could be quickly sent a mile or two to the firehouse. Even before the fire alarm system a built-up part of a city always had a fire station within a few blocks, but in a town strung out for two miles along Main Street the news of a fire took some time to reach someone near the firehouse.

Furthermore, the electric siren has a far greater carrying power than the old fire bell or locomotive tire struck by a sledge hammer. Also, someone had to get to the firehouse and perhaps wait for a key before he could pull the bell rope. Today companies which have no paid men in the firehouse arrange with someone living nearby to take fire calls and punch a siren button located in the house. Thus women fire dispatchers have become ubiquitous in volunteer companies.

It is significant that a sampling of such companies across the nation shows that well over half have been established since 1910.

A Pennsylvania study by Elizabeth Smedley in 1962 reported the formation of 20 new volunteer companies a year. The Montana Fire Service Academy reports an average of two new companies a year in that state. Other states range in between. This is, of course, partly balanced by large, growing communities shifting to paid departments. For instance the chief of the Harrisburg, Pennsylvania Fire Department reports the almost total disappearance of city volunteers due to their shift to suburban companies. However, some suburban volunteers also serve in bordering city areas.

As will appear, the modern volunteer fire fighter is much better trained than firemen of the past. For one thing the fire fighters today use much more complex and sophisticated equipment. Thus Robert S. Holzman in his history of fire fighting (1956) could say, "Today many volunteer companies have apparatus that would shame most professional departments." On the basis of her survey of Pennsylvania fire companies Elizabeth Smedley stated, "the equipment of paid companies is sometimes inadequate and obsolescent while volunteer companies are often much better equipped." This is in accord with John V. Morris's statement of 1955 in his history of fire fighting: "Up-to-date equipment may be found on rural apparatus long before it is adopted by big cities, because of the red tape and complicated training program involved in introducing revolutionary devices and methods to the larger departments." More to the point is Smedley's explanation: a large part of the budget for paid departments is expended on personnel.

For instance, the small Gladwyne, Pennsylvania department founded in 1944 got high pressure fog equipment in 1954 before it was adopted by Philadelphia. (High pressure fog is produced by a pump with a third stage delivering up to 600 pounds pressure, and a special nozzle. It permits fire fighters to enter a hot room behind an insulating curtain of fog spray and often puts out a fire with little water damage.)

The following chart shows typical kinds of equipment owned by fire companies in rural areas, small towns, and small cities.

COMMUNITY	POPULATION	EQUIPMENT
Sandy RFD, Oreg.	4,180	3 — pumpers 4 — tankers 2 — rescue
Lakeside RFD, Oreg.	1,422	2 — pumpers 1 — tanker 1 — rescue

COMMUNITY	POPULATION	EQUIPMENT
Clifton, Wis.	3,300	2 — 750 g.p.m. pumpers 2 — 500 g.p.m. pumpers 1 — 65' aerial ladder 1 — jeep
Bowie, Tex.	5,185	1 — 1,000 g.p.m. pumper 1 — 750 g.p.m. pumper 1 — 500 g.p.m. pumper 1 — pickup
Ketchikan, Alaska	6,994	1 — 1,500 g.p.m. pumper 1 — 1,250 g.p.m. pumper 1 — 750 g.p.m. pumper 1 — 75' aerial ladder 1 — 80' snorkel 1 — fireboat
Centerville, Iowa	6,531	1 — 750 g.p.m. pumper 1 — 600 g.p.m. pumper 1 — 65' aerial ladder
Jamestown, N. Dak.	10,601	1 — rescue 3 — pumpers 2 — 1,000 gal. tankers 1 — chief's car 1 — station wagon
Martinsville, Va.	19,653	1 — 1,200 g.p.m. pumper 1 — 1,000 g.p.m. pumper 1 — 750 g.p.m. pumper 1 — 500 g.p.m. pumper 1 — 75' aerial ladder 1 — salvage unit 1 — officer's wagon
Franklin Park, Ill.	20,348	3 — 1,250 g.p.m. pumpers 1 — 85' snorkel 1 — 100' ladder

COMMUNITY	POPULATION	EQUIPMENT
Florence, S.C.	25,997	3 — 1,000 g.p.m. pumpers 1 — 750 g.p.m. pumper 1 — 350 g.p.m. pumper
Monroeville, Pa.	29,011	1 — 1,500 g.p.m. pumper 1 — 1,250 g.p.m. pumper 1 — 85′ snorkel 1 — rescue 1 — chief's car 1 — ambulance
Cape May Point, N.J.	325	1 — 1000 g.p.m. pumper 1 — 600 g.p.m. pumper 1 — utility pickup

BASIC EQUIPMENT ON EACH PUMPER

2 way radio and walkie-talkie
2000′ 2.5 or 3″ hose with 3 or 4 nozzles
500′ 1.5″ hose with 2 nozzles
500′ to 1000′ booster hose (i.e. that is attached to tank) with 2 nozzles
16′ hard suction hose
6′ "soft suction" hose (for hydrant rather than suction:)
2 gates for hydrant
2 hydrant wrenches
1 or 2 "Y's" or Siamese to connect 2 smaller lines to a larger one
1 35′ extension ladder
1 12′ ladder
1 folding roof ladder
1 4′ cellar pipe with nozzle
6 or 8 spanners for hose connections

10 or 12 adaptors to connect large hose to small or join mismatched ends

2 axes

1 or 2 prybars

2 shovels

2 or 3 pike poles to pierce ceilings or pull down burning woodwork

1 bolt or chain cutter

4 to 8 plastic or canvas tarps to cover furniture

4 to 6 Air Pacs (self-containers breathing apparatus)

several kind of rope

strap and asbestos gloves for hot appliances

weighted chimney mop or Chimflex (a torch which exhausts the oxygen in a chimney)

bag of salt for icing conditions

first aid kit

3 or 4 battery hand lamps

1 CO_2 extinguisher

1 dry powder extinguisher

1 pressurized water or soda and acid extinguisher

ADDITIONAL ITEMS ON OTHER ENGINES OR PICKUP TRUCK

stationary or portable generator with flood lights

1 or 2 smoke ejectors

portable pumping unit for small streams

Halligan tool for forcible entry

4 or more Indian tanks (5 gal. shoulder tanks with hand pumps for field and brush fires)

several 5 gal. cans of foam

foam mixer

foam nozzle

wooden or metal ramps to permit vehicles to cross hose

wheel blocks
electric deodorizer for smoke-filled house
water vac to get water out of building
electric saw
oxygen resuscitator
hose clamp to shut off broken hose
hose jacket for break
tow chain
battery jump cable
electric coffee maker
jug of drinking water

Equipment varies according to location conditions. A rural company will probably not carry foam equipment but will have numerous Indian tanks and beaters, whereas an urban company may have few or none. A department with a boat will have life jackets and grappling equipment, etc. A company with an aerial ladder or Snorkel will have men specially trained for such equipment. As mentioned, rural companies often have tank trucks equipped with pumps to supply engines.

The age of the equipment ranged from the 1950s to 1977, usually spread out among the vehicles. As 20 years is normally considered to be the efficient life of a pumper, it is clear that most of the pieces listed fall within the projected life span. The older vehicles are ordinarily used for reserve or back-up purposes. Furthermore, a company in a small community which has few fire calls is likely to find that the apparatus has a longer useful life than that in very active companies. In addition, fire apparatus is regularly inspected and tested by the underwriters, who determine insurance rates in each community.

This brings up the question of finance. Where do volunteer companies get the money? The answer is from various sources. Responses from about 45 companies across the country show that 19 are entirely supported by municipalities or regional authorities; another six by contributions, raffles, bingo games, etc.; and 16 by both governmental and private sources. But when it comes to the ownership of apparatus, in 31 communities equipment is publicly owned; in nine it is owned by the fire companies; and in six, owned jointly.

This repeats a pattern going back to the earliest days. At first the fire companies used municipally-owned apparatus, then bought their own, but often asked for municipal financial aid. Municipalities tended to buy the first steamers. For the same reason there has been a recent shift to municipally-owned apparatus. In the days of the hand tub, a company might raise enough by subscriptions to pay for it, and one might be used for 50 or 60 years. Steamers were, of course, expensive and so is modern apparatus. A pumper that cost $35,000 in 1960 probably costs $80,000 today. The early aerial ladders of the 1920s cost about $18,000; today a ladder truck or a snorkel costs at least $150,000. That kind of money can't be raised by dues, raffles, or bingo. And of course the cost of the engine is only part of it; a pumper may carry $5,000 worth of hose plus thousands of dollars worth of other equipment.

Although the municipality may own the apparatus, other equipment may be paid for out of dues, contributions, and fund-raising activities. It will be noted that more than half of the companies surveyed obtained money from other sources. This is in accord with the 1962 findings of Elizabeth Smedley in her study of Pennsylvania fire companies.

When public support is inadequate, volunteer firemen sometimes build their own apparatus. The Eleanor Volunteer Fire Department in West Virginia founded in 1963 in an unincorporated town received only $500 a year from the county; therefore the members of the small department built their first pumper from a 1956 Ford truck. In 1976 it was still in service along with a newer pumper. Similarly, the newly formed Hobe Sound Fire Company in Florida built their first piece of apparatus.

Small-town and rural companies which operate in areas without piped water often get hold of a second-hand tank truck and convert it for fire service. Firemen in small places sometimes build or help to build their own firehouses. As a rule all these activities are aided by contributions from the community and the fire fighters themselves.

Fund raising also ties in with the social activities of a volunteer company. Bingo and raffles are social functions as well as money-raising projects. Various other money-making schemes are common: flea markets, cake sales, subscription suppers, carnivals, and services such as pumping out cellars or burning off vacant land. In many of these the women's auxiliaries participate. One company holds a three-weeks-long Christmas tree sale, bringing in a profit of $4,000 to $6,000. One or another of the firemen's wives makes up scores of decorated wreaths. Part of the proceeds go for a cocktail party or dinner dance for the workers and their spouses.

Many companies use the parade as part of a fund-raising activity. In Northeastern Pennsylvania, an area served by 166 volunteer fire companies, there will be a fire fighters' parade on almost any weekend from early June to late August. Of these companies, all members of the Northeastern Volunteer Firemen's Federation, 159 are almost entirely self-supporting. Thus each parade is planned to attract the largest possible crowds. Neighboring fire companies, marching bands, drill teams, fife and bugle corps, and local dignitaries are invited to participate. The parade will wind up at a picnic or bazaar offering food, drink, rides, raffles, bingo, and fire fighters' competitions.

One of the most successful of these is the carnival held by the Clifford Township Volunteer Fire Company, which in 1957 had the foresight to buy 22 acres of flatland which it turned into a fairground. At its annual four-day carnival there it takes in between $50,000 and $75,000. In a township with a population of about 15,000 the carnival attracts nightly crowds estimated at over 10,000. As a result the company does not need to get municipal funds or solicit contributions. Yet it is able to maintain eight vehicles including an ambulance and an antique 1922 chain-drive Mack ladder truck used for kiddie rides.

As John Uram, publisher of the *Northeast Volunteer* says, "Volunteer firemen spend most of their time raising money." The Taylor No. 1 Fire Company, which went $120,000 in debt to build a new firehouse, was able to pay off $90,000 in five years largely by catering banquets in the firehouse—about 26 weekend bookings a year.

Not all companies are so successful. The Lake Silkworth Volunteer Fire Department threatened to close its doors unless the township increased its annual appropriation of $1,000. The township added another $1,000 and the company grumblingly stayed in business.

Such appropriations in small municipalities are not unusual but are grossly inadequate. Modern fire fighting gear is expensive. Fire hose costs $2.60 a foot for the 2½-inch size; $1.95 for the 1½-inch. A pumper may carry 2,000 feet of the larger size; 400 of the smaller. In rural areas where water must be pumped from a stream or pond or in suburban communities with widely-spaced hydrants a hose lay of 3,000 to 4,000 feet may be required. As for nozzles, the 1978 price is $200 each.

Turnout gear: coat with detachable winter liner, helmet, boots, and gloves costs about $150; thus a volunteer company with 45 fire fighters must invest $6,000 to $7,000 in clothing alone—not all at one time, of course. An established company has bought its outfits over a period of years.

Smoke inhalation was always a hazard, but with the increasing size of apartment buildings and the ubiquity of toxic fumes (for instance those given off by plastic items in drug and variety stores) the use of air paks has become nearly universal. (These are air tanks feeding into a face mask.) Even a small company may own half a dozen at a cost of $450 each.

For instance, the budget of the Goshen, Pennsylvania Fire Company with 50 volunteer fire fighters is as follows:

INCOME:

$22,000	County fair
22,500	Letter drive
13,500	Municipal donations
16,500	Bingo, dinners, etc. and money distributed to volunteer fire departments by the Pennsylvania Dept. of Revenue from the 2 percent tax on out-of-state insurance companies.
$74,500	

EXPENSES:

$ 3,975	Maintenance of building and grounds
7,425	New equipment, hose, etc.
8,475	Maintenance of apparatus and communications system

6,000   Fuel for trucks, oxygen, & supplies

6,150   Heating oil, electricity, etc.

9,975   General insurance on building, trucks, etc.

3,525   Operating disbursements: dress uniforms (members must pay back cost), welfare (flowers, cards, etc. & cost of letter drive).

4,500   Special disbursements from relief association from funds from the Pa. Dept. of Revenue, used for breathing apparatus, portable radios, turn-out gear, insurance for members.

24,750   Reserve for new apparatus

$74,775

No other public service is so likely to be entirely or largely self-supporting as a volunteer fire company. The police, trash collection, and street maintenance are all paid for by taxes; the customer pays for water, gas, electricity, and often for sewers; hospitals, in addition to public funds and contributions, charge for their services, but no one is charged a fee for a call to a volunteer fire company.

In view of the amount of money involved, it is obvious that running a volunteer fire company requires some sophisticated financing and accounting. These finances are usually handled by a set of officers other than the chief (often called a chief engineer) and other fire officers. Thus there will be a company president, a secretary, a treasurer, and often a board of directors. These officials are not necessarily fire fighters, although a number of them may have been so at one time. Theoretically the chief and the assistant chiefs are responsible to the president and directors, but except for the handling of finances, the brigade officers pretty much run the company, subject to approval by the directors. For one thing the fire officers normally enjoy the loyalty of the active fire fighters, and in case of controversy, their support.

In addition to drilling, fire fighting, maintenance of apparatus and house, plus fund raising, the active members as a rule organize the social functions of a fire company. It is customary to invite other companies to certain events such as housings, dedications, and regular get-togethers.

These social activities have great value in developing mutual acquaintanceships which pay off at fires involving two or more neighboring companies. Today mutual assistance agreements are the rule. Every department responding to the aforementioned questionnaire had such a pact with neighboring communities.

The state of Oregon has one of the most structured systems: a local chief may call upon the county fire chief for assistance, who will dispatch mutual aid equipment. The county chief may, if necessary, call upon the district chief or in his absence upon the state fire chief. In a grave emergency the state chief can advise the governor of the need to invoke the provisions of the Emergency Conflagration Act.

The Oregon State Fire Marshal's office has statistics on the apparatus owned by each company, the number of men available, the number of alarms answered, and the fire losses during the proceeding year. On the other hand, a 1975 study of fire service in Connecticut stated that "Meaningful statistics for the State of Connecticut are virtually nonexistent. . . . No one knows how many pieces of fire apparatus exist in the state, how many active fire fighters there are, or even how many fire departments." Recognizing that "most of our municipalities still have volunteer fire departments," the survey committee proposed that the position of State Fire Administrator be established to coordinate fire protection.

Other states, for instance, Michigan, North Dakota, Pennsylvania, and Virginia, publish the names and addresses of all fire departments, but not the number of fire fighters available nor a listing of apparatus. Most states apparently have no meaningful statistics on fire protection, nor any overall plan.

The Oregon plan was clearly designed for a region with much forest land and many widely separated communities. In relatively metropolitan areas such as suburban townships adjoining Philadelphia, the township fire marshals dispatch the local volunteer companies and take command when more than two companies are involved. Otherwise the chief of the first company to arrive takes charge. In a more sparsely populated area such as Chester County, which sprawls over 760 square miles, has 57 townships, 15 boroughs, one city, and 45 volunteer fire companies, it was recognized that county coordination was essential. Two groups of five fire companies asked the county commissioners to establish a fire board. Fire Chief John H. Dean, a leading proponent of the plan, was tapped for the job. He immediately started correlating data from the 45 volunteer companies, working out dispatch procedures, and assigning territories.

On October 2, 1971 the plan went into effect. The heart of the system is a console that controls two multifrequency radio stations six miles apart. All messages and telephone calls into and out of the center are recorded and can be played back. A microfilm display on the console contains information about streets, route numbers, building floor plans, as well as a map of the county showing the operational area for each company. The dispatcher then sends the appropriate companies and ambulances, and if any area is depleted, puts move-up plans for mutual aid into action.

In less-structured areas a disregard for political boundaries is a natural characteristic of volunteer companies. As the Connecticut study shows, they tend to be largely self-governing, and the volunteer instinctively pitches in when he hears of a fire. A few years ago suburban Philadelphia firemen driving through South Jersey stopped off to fight forest fires 50 miles from home.

But whether mutual aid is based on informal agreements or on more centralized structural arrangements, modern volunteer fire companies are not a collection of warring tribes but are highly cooperative organizations. In suburban areas it is not unusual to have six or eight companies working together on the equivalent of a city multi-alarm fire. And unlike local police forces with no authority outside a city or township, volunteer fire companies cross borough, township, county, or even state lines when called upon for assistance. Pennsylvania companies cross into New Jersey, New York, Maryland, and West Virginia.

At the Garden State Race Track fire in Cherry Hill, New Jersey, April 14, 1977, there were 60 engine companies, 16 ladder companies, and 28 ambulances or rescue vehicles. Included in the 104 pieces of apparatus were five from Philadelphia. Altogether 700 volunteers participated, safely evacuating 12,000 people, including lucky winners who were reluctant to leave before cashing their tickets.

Forest and brush fires are no respecters of district or county lines. In 1977 such fires devastated over 40,000 acres in New Jersey, 850,000 in Alaska. In August fires raged over 276,745 acres in California. One near Santa Barbara destroyed over 200 homes. In Maine one fire required 15 days to extinguish; another in New Jersey took 13 days; and the Marble-Cone fire in the Big Sur country in California took over 16 days and required nearly 6,000 fire fighters from 23 states.

Most of these were regular employees of state and federal

forestry agencies, but others were students and men willing to work for $3.90 to $4.50 an hour plus 25 percent for hazardous duty. Obviously such a fire was far beyond the capacity of the Big Sur Volunteer Fire Brigade of 15 men serving a region 30 miles long. Some of their equipment came from donations or the repair of discarded damaged pieces. Eight volunteers have fast-attack slip-on pumper units on their four-wheel drive trucks.

Fighting forest and brush fires is exhausting and at times dangerous work. Weary men without much rest may be called out for another fire. In July 1977 four men ranging in age from 23 to 52 from the Eagleswood Volunteer Fire Company No. 1 died when trapped with their apparatus in the Bass River State Forest near Tuckerton, New Jersey. They were not even in their own county.

At their funeral several thousand firemen from four states rode 137 fire trucks in the procession. The bodies were carried on Eagleswood's brand new pumper and on the company's first engine bought in 1927. Returning from the funeral, volunteers were called out to join with Forest Service men in fighting another blaze in the area where the men had died.

Of course, mutual aid in conflagrations is an old tradition. The response of companies from as far away as Philadelphia to New York's great fire of 1835 has been described. When Boston had a great fire in November 1872 companies came from Cambridge, Brookline, Newton, Lynn, Salem, Waltham, and from as far away as Fall River, Providence, Rhode Island, and Portsmouth, New Hampshire. Some of these were no doubt volunteer outfits. One old hand tub from Wakefield 12 miles away came and went to work. Baltimore's great fire of 1904 brought engines from Washington and Philadelphia, transported on flat cars.

However, these were unusual occasions; whereas the routine response of Pennsylvania companies across state lines is rather exceptional. Nor are professional and volunteer companies mutually exclusive. City companies assist in the suburbs and vice versa. On a night in March 1963 during a howling gale Terre Haute, Indiana had a downtown fire which required every company in the city. The paid department is an excellent one, but at 3 A.M. Chief Fesler radioed for help from volunteer companies in surrounding communities. A dozen responded from as far as 80 miles away. Afterward, Chief Fesler said, "Without the help of the volunteers we probably would have lost our entire downtown."

However, the Philadelphia Fire Department uses an uncommon type of hydrant and pumper connections so that they cannot be hooked up with suburban apparatus without adaptors. The engine fittings are the same, however, so that one place can relay to another.

One important difference between professional and volunteer companies is that the latter cannot specialize to the same degree. As the listing of equipment owned by the volunteer companies shows, they do not break down into engine companies, ladder companies, rescue units, and ambulance service. Some of them also have boats for water rescue. The ambulance crew may be somewhat separate from the fire fighters, but because of the uncertain composition of a group answering an alarm they cannot be previously differentiated into ladder men, hose men, and rescue crew. Men and women joining a volunteer company are trained in all three departments. Only the drivers and pump and aerial ladder operators receive specialized training, and if they are not needed for those functions on occasion, they may act as hose or ladder men.

A suburban volunteer company has to be prepared for a wide variety of situations: everything from woodland fires to those in high-rise apartments, and in many places, industrial fires. Pointing to the migration of industry to suburban and rural areas, the governor of Oregon issued an executive order in 1974, setting up a State Fire Service Plan to coordinate the work of local companies at industrial fires.

Suburban and rural highways are thronged with trucks carrying all sorts of hazardous materials. A rural company may be faced with a blazing oil truck or one loaded with poisonous chemicals.

With the decay of railroads, derailments of tank cars loaded with dangerous materials have become common. These, of course, may occur anywhere, but are often in areas served by volunteer fire departments. On the weekend of February 25, 1978 a propane car exploded in the center of Waverly, Tennessee, a town of about 4,500 people. The blast and fire killed 15 people including the local fire chief, Wilburn York, and destroyed or damaged 14 buildings. Two thousand people had to be evacuated. That same weekend a train jumped off the rails on a trestle near Cades, 80 miles west of Waverly. There were no injuries or fire, but one of the derailed cars carried sodium hydroxide, a caustic lye. And while the propane was being pumped out in Waverly, 17 cars derailed near Youngstown, Florida. Six of them carried chlorine, and one of them ruptured, spreading the gas over a sleeping farming community and killing eight persons.

To make things worse for local fire companies and rescue squads, the railroads' chain of command is so complicated that often no one knows who is in charge. In the Waverly situation it took eight hours to learn that the derailed tank car was filled with propane. It lay on the track for two days before a crew came to empty it. Despite the danger, crowds came to watch the operation. Local police and firemen were standing by in case of fire, but the blast, when it came, devastated an area the length of two football fields.

Two pictures of *Corinthos* tanker fire, Marcus Hood, Pennsylvania, January 13, 1971. (The Evening Bulletin, *Philadelphia*)

Volunteer firemen fighting West Conshohocken, Pa. blaze, which was fed by a broken gas main and destroyed 20 homes, January 21, 1971. (The Evening Bulletin, *Philadelphia*)

Ice-covered pumper after battling the gas main fire for more than three hours (The Evening Bulletin, *Philadelphia*)

A seventeen-year-old volunteer fireman. Her brother is one, too. (The Evening Bulletin, *Philadelphia*)

Women fire fighter passes test in Massachusetts (*UPI photo*)

A typical training tower

Old engine but modern membership, Crescent, Oregon.

Fire companies and civil defense workers poured in from surrounding communities. In addition to the volunteer personnel, Nashville, 60 miles away, sent police, ambulances, civil defense units with portable generators, and two foam trucks plus a light truck with a generator. It is, of course, traditional for volunteer and professional units to work together in a major disaster. But perhaps more often than not volunteer companies must be first on the scene at highway and rail catastrophies.

Most volunteers have had some instruction on how to handle dangerous cargoes, but small-town and rural companies cannot afford some of the necessary specialized equipment such as foam trucks and large portable generators. For one thing, although railroads and trucking companies often need the services of volunteer companies, these corporations rarely contribute for their equipment or upkeep. The most a small-town company can expect is a token contribution for long and often dangerous service at a truck crash or a derailment. A volunteer department is lucky to be able to persuade a $200,000-a-year executive to authorize a $200 payment for 40 or 50 man hours and the use of expensive equipment to handle a truck or freight car wreck or fire.

To a great degree each small town or suburban unit must be complete in itself. Thus, in addition to pumpers and ladder trucks, a company may operate a field piece—usually a four-wheel drive vehicle with a water tank and portable pump. It will be noted that rural companies commonly have large tank trucks.

But vehicles are only part of the story. A modern fire company carries scores, perhaps hundreds of specialized items: foam equipment, shoulder tanks for field fires, electric generators to run smoke ejectors and flood lights, electric roof saws, bolt cutters, specialized entry tools, tarpaulins to cover furniture or holes in a roof, along with a wide variety of connectors: T's, hydrant gates, adaptors for different hose size or to join two male or two female fittings. (Hose laid from a hydrant comes off the engine ready to connect at each end, but a reverse lay requires an adaptor at each end. A garden hose can easily be reversed if the wrong end is left at the faucet but several hundred feet of fire hose cannot be—thus the need for adaptors.) A missing hydrant wrench can be a disaster, so all companies carry spares. A stuck connection requires special spanners to separate it. Men must learn where all these things are kept.

The use of foam to blanket oil fires requires special equipment: an inductor and a nozzle extension. Three or four types of extinguishers are ordinarily carried for quick use on small fires: soda and acid, pressurized water, dry powder, and CO^2. Each has its preferred use.

If a company acquires an articulated boom, a special training program must be instituted. The Snorkel Company, for instance, furnishes a film showing the numerous steps in the operation of the equipment plus a variety of cautionary admonitions. It also supplies a 33-page set of instructions which the members of a company must study. On top of this the manufacturer sends instructors to teach the fire fighters the intricacies of the equipment. For it is a serious matter to operate a $190,000 machine which puts men 65 to 85 feet in the air, and can be used for rescues at those heights. Also there are dozens of maintenance procedures.

The hazards include coming in contact with electric lines, setting the outriggers on unstable ground, extending the boom too far to one side, and overloading the platform—not to speak of possible mechanical failures in one or more of the many complex components. There are, of course, fail-safe features which the operators must learn. Even a person on the ground who touches a machine in contact with a live wire can be electrocuted.*

Before and after use the machine requires maintenance procedures. Between alarms the company must schedule regular practice in these.

At present only a small proportion of volunteer companies have articulated booms, but the number is increasing as office buildings and high-rise apartments encroach into areas beyond city boundaries. More common is the aerial ladder truck, which also requires technological knowledge to operate and has its own hazards. Even the hoseman at the top has to be trained in proper procedures.

Long gone are the days when a suburban company was concerned only with brush and small building fires. As a picture of an Ardmore fire in 1977 shows, even a four-story office building requires the use of either a boom or an aerial ladder. Another important item at a large fire is the deluge gun, which was also used. This is a kind of water cannon fed by two or more large hose lines, and often requiring the use of more than one pumper. Thus it can involve two or three fire companies. Five volunteer companies fought the blaze in Ardmore, which threatened much of the business district.

Even at a house fire an aerial ladder or a boom is useful because it enables a fire fighter to pour water on the blaze from above; for an apartment fire such a piece of equipment is a must. The usual portable ladder extends to only 35 feet, and even at that height is not very satisfactory for rescue work, for it has no side rails. The aerial ladder with its side rails and large treads permits a series of people to descend, providing they are reasonably agile. Although the boom can handle fewer people at a time, its advantage is that even the infirm can step onto the platform or be lifted onto it.

---

* Like motorists in a car touching a wire, people must avoid simultaneous contact with the machine and the ground.

Thus at an apartment fire of any size it is desirable to have both types of apparatus on hand, which normally means the participation of two or more fire companies. (Only a rare volunteer company has more than one aerial piece.) Here again is a reason for cooperation among neighboring fire companies—a cooperation requiring pre-planning and occasional joint drills.

All this shows that membership in a volunteer company involves far more than appearance at an occasional fire: it requires training, often highly technical, and it means many hours of rather dull work. It is exciting to lay out hose and attack a fire; it is much less glamorous to go through the routines of training sessions. But rigorous training has become a part of the life of a volunteer fire fighter.

Long past are the days when it was enough for a fireman to have big muscles, a hard head, and to be sober enough to toss a bucket of water or pump an engine.

Thus all sorts of training programs have been set up, ranging from the company drills to state fire schools. Almost every company holds drills one to four times a month. As a rule these are supervised by experienced officers such as former professional firemen, or most often by men who have attended a fire school. These are operated by townships, counties, and states. Today every state in the union operates one or more fire schools, usually under the jurisdiction of the State Department of Education and often connected with a university.

Nebraska's first three-day fire school began in 1937 at the instigation of volunteer firemen in the state. In Pennsylvania the first one was probably that organized in 1932 by 15 members of Reading's Keystone Hook and Ladder No. 1. Long before such a requirement became common the company recommended that all new applicants for membership take a six-month's training course.

A state-sponsored school in Pennsylvania was begun in 1939 at State College, then moved to Lewistown four years later. In 1949 the legislature established a permanent Firemen's Training School under the control of the Department of Public Instruction. All publicly employed firemen and all enrolled members of volunteer companies are eligible for admission. The Public Service Board apportions the number to be admitted on the basis of the population of various counties.

All state fire schools offer basic and advanced courses. For instance, Virginia offers certificates for courses of 30, 60, and 120 hours, open to both paid fire fighters and volunteers. Starting with 80 trainees in 1966 the program had 175 in 1969 and 400 in 1975. At the Iowa State Fire School in 1974-1975 the basic program had 975 enrollees representing 400 communities. The total enrollment that year for all programs was 6,115, representing 91,700 man hours of instruction.

In New Hampshire's first full-time recruit school course of eight weeks—longer than most—22 percent of the enrollees were volunteers. Michigan, which in 1971 had four schools offering a 66-hour course, and 204 enrollees, by 1974 had 54 schools offering 66-hour courses and 68 with 240-hour courses, and combined enrollments of 4,802. The courses are open to both paid fire fighters and volunteers.

Typical courses in a fire school cover such subjects as chemistry of fire, fire hose practices, ladder work, pump operation, ventilation, small tools and forcible entry, salvage, and rescue. For officers there are advanced courses in fire department management, arson detection, fire ground tactics, fire fighting during civil disturbances, and, of course, pre-planning.

Along with courses in theory the fire fighters get intensive practical experience. A fire school usually has a four-story tower in which fires are built. Trainees using air paks (breath-

ing apparatus) must enter a smoke-filled basement or set up
ladders and carry up hose. There is also a pit for oil and
gasoline fires for practice with high pressure streams or foam.

A number of states have set up their fire schools on a
regional basis. County and township schools tend to be limited to basic training, offering 15- or 20-hour programs.

In any case a majority of volunteer companies responding
to a questionnaire state that new members must go through
probationary training. Usually this means participation in
the company drills, but increasingly new members are required to attend a fire school.

The twentieth-century volunteer fire company is therefore
very much a part of a technological revolution. The introduction of motorized apparatus, the telephone, and more
recently radio made the modern volunteer company possible; it enabled rural and scattered communities to establish
effective fire companies. In the past there was no point in
setting up such a company because men could not get to the
firehouse in a hurry or pull an engine a couple of miles in
time to be of any use.

Now fire fighters, even those beyond the reach of a siren,
often have radio-controlled alerting units in their homes,
their cars, and at work. Using their cars they can be at the
firehouse or the fire in minutes and an engine can quickly
travel miles. Big tank trucks carry water to places beyond
the mains.

By radio the first officer to arrive on the scene can direct
the placement of incoming engines and tell the crew when
and where to lay hose, and what equipment to use. At a fire
covering a large area such as woodland, companies can keep
in touch by radios on the trucks and by walkie-talkies for the
men on foot.

All this revolution in equipment has restored the importance of the volunteer fire company and has caused it to
become technically skilled.

IX

MEMBERSHIP

THE EARLY FIRE COMPANIES, as noted in earlier chapters, drew their membership from among leading citizens and respectable artisans. This pattern continued as population moved westward; thus at a time when fire companies in Eastern cities had fallen into the hands of rowdies, those in places like Reading, Pennsylvania, New Orleans, Milwaukee, Chicago, and even San Francisco, where respectable volunteers were not plentiful, tended to attract prominent and solid citizens.

Today the volunteer fire company is rarely made up of rowdies. The social classes represented vary depending on tradition and the nature of the community. Thus it is obvious that the fire company in a predominantly working-class town will be drawn from that class. In a small city or middle-class suburb the companies will tend to draw on a more varied personnel. A well-to-do suburb is likely to have a number of college students in the fire company along with a sprinkling of business and professional men, and increasingly women.

On one occasion in Gladwyne the crew of the second piece out consisted of a dinner-jacketed man who had come from his daughter's debutante party, the local garage man, a lieutenant commander in the Navy, and a college professor of English. In answer to a Pennsylvania questionnaire in the 1960s, 21 of the 38 chiefs of volunteer companies stated that the average working man makes the best fireman; eight preferred skilled workmen, whereas the chief in a large Philadelphia suburban township said, "We have them from all walks of life, rich and poor, doctors and laborers."

It is small wonder that the pedantic anthropologist in John P. Marquand's *Point of No Return* is baffled by the social structure of a firemen's muster in Clyde [Newburyport], Massachusetts.

"And, my God, this thing,"—Malcolm Bryant waved his arm in a gesture that embraced the training field—"this beautiful, tribal ritual. It's like a Maori war dance. I'm just beginning to get it straight. It doesn't include the whole tribe does it?"

Discovering that John Gray, a Harvard man of the lower-upper class, is up there on the engine, the anthropologist marvels "And he's up there. . . . Now that's very interesting. Up there and out of his group. It's going to take me quite a while to get this structure straight."

More important than social class are certain other qualifications: age, physical fitness, character, standing in the community, civic-mindedness, and until recently race and sex. Nowadays a medical examination is often a requirement for membership. Minimum age requirements usually vary between 18 and 21, with some companies operating a junior program for 16- to 18-year-olds. In areas where a high proportion of young people are in college, the training of 16- to 18-year-olds is useful because such persons may later join as regular trained fire fighters. Maximum age for joining tends to be 40 to 55, with an upper limit of membership of 55, 65, or none at all.

In the suburbs populated by commuters, the fire fighters available for a daytime alarm are likely to have some retirees among them. Because of the daytime exodus a number of suburban companies are accepting women members. At a fire the younger personnel wear the air paks and climb ladders; the older men hook up and pack hose, cover furniture with tarpaulins, and assist with the lighter chores. In one New Jersey resort village with many retired citizens, perhaps 50 percent of the fire company is over 65 years old. Of the 25 active members of Washington Engine Company No. 2, Riverhead, New York in 1976 nine had joined between 1935 and 1947.

Far from preferring the daredevil, the chiefs of 116 volunteer companies in Pennsylvania described the best fireman as having a calm and cautious nature. They used such phrases as "level-headed," "not easily excited," "calm," "cool-headed." By contrast, "courage" was listed by only two chiefs of predominantly paid companies and 22 volunteer companies.

On the other hand, the chiefs of 99 volunteer companies listed civic-mindedness or interest in the community as an important characteristic of a good fireman. One reason is that in many small towns the fire company takes responsibility for such community activities as fund drives, sponsorship of Boy and Girl Scout troops, aid to families stricken by disaster, blood donations, and, of course, ambulance service.

Comparative statistics on ambulance service are not very enlightening because of the wide variety of practice. In some communities the ambulance unit is not part of the fire company; in others the police operate the ambulance service, and a "rescue" vehicle may or may not be an ambulance. The last figures available for Pennsylvania show an increase of companies providing ambulance service from 11 percent in 1946 to 26 percent in 1957. In Oregon, with a large number of rural companies, approximately 17 percent of all those reporting in 1975 had ambulances—some as many as three. Of the 740 volunteer departments in Minnesota 145, or nearly 20 percent, run ambulances.

The volunteer fire department is, therefore, a social agency and often very much a part of community life. This has been true over the centuries.

Inevitably, a close-knit organization like a fire company, and one involved in community affairs, gets political power. The bylaws of the early companies sometimes prohibited political discussion at company meetings but obviously such a prohibition would not prevent unofficial discussion and action. From very early times office-holders and candidates found a useful base in a fire company.

In 1771 the entire City Corporation of Philadelphia seems to have belonged to the Hand-in-Hand Company. Six early mayors of that city were volunteer firemen. In New York, as mentioned before, William Tweed used the Americus Company as a political base. As Alexander B. Callow, Jr. wrote in his history of the Tweed Ring, "Tweed led an unusually cohesive company of 75 men who could be counted on to vote the Tweed scriptures on issues and candidates and lend a bully's hand on election day." Tammany Hall had as its symbol the tiger from Tweed's engine.

The political rise of David C. Broderick, a more estimable character, has been mentioned. From New York, where he had been foreman of Engine 34, he came to San Francisco in 1849 and immediately organized the first engine company,

Empire No. 1. Using this as a political base he was elected to the state senate in less than a year, became its president in 1851 and U.S. senator in 1856.

Today the mayor of a small town almost certainly is or has been a member of the local fire company.

The political influence of the volunteers kept New York from having a paid department until 1865 and Philadelphia in 1871, long after Boston, 1837; Cincinnati, 1852; Chicago, 1858; and Baltimore, 1859. Firemen in Reading, Pennsylvania defeated attempts to establish a paid department in 1914, 1926, and 1932.

The cohesiveness of a volunteer fire company is to some extent due to the fact that from early times to the present it elects its own members. Thus new members are usually known to at least some of the company and are likely to be from similar social backgrounds. This has both good and bad effects. It permits a well-run company to keep out undesirables or to drop someone whose behavior is objectionable.

One type of undesirable against which the volunteer company must be on guard is the firebug. It is understandable that someone with a neurotic compulsion to set fires should be attracted to a fire company. For some arsonists the compulsive lure of fire may be less of an incentive than the excitement of answering an alarm, an emotion shared by the spectators who jump in their cars to follow the engines, tie up traffic, and generally get in the way. But whether the firebug acts out of boredom or a sexual compulsion, he occasionally turns up in a fire company.

Most often the gung-ho firebugs confine themselves to setting brush fires for the excitement of riding the engine, but the really dangerous type will set off vacant buildings or even places closed for the night. In the 1920s some volunteer firemen in Harrisburg, Pennsylvania were finally caught after they had set fire to a furniture warehouse and a lumberyard. In May 1976 a volunteer fireman was charged with 16 counts of arson and one count of homicide in connection with fires in East Lansdowne, a suburb of Philadelphia. A former fire chief there was convicted of covering up evidence against the man. In Shamokin, Pennsylvania two firemen were charged with setting a motel fire that killed nine people in January 1977.

Minor incidents such as incendiary brush fires may never be reported in the newspapers or may not reach a magistrate. The fire company may simply expel the culprit or, in the case of a rookie, suspend him. A volunteer company quickly becomes suspicious of a member who is too often hanging about just before alarms for fires of dubious origin. Furthermore, with arson, which accounts for as many as 30 percent of all building fires and is usually for profit, volunteers are increasingly enrolling in courses of arson detection.

Two other vulnerable areas for volunteer companies are mentioned by a writer in *Fire Command* for July 1970. Although on the whole favorable to volunteer companies, he mentioned as a weakness a tendency to select officers on the basis of personality rather than for qualifications and leadership. Aside from the fact that personality is an important factor in leadership, it should be noted that paid departments are not always free from cronyism and political influence. For instance, in Philadelphia the mayor appointed as Fire Commissioner for the city his own brother, a man whose qualifications have come under question, especially after a refinery fire in which nine firemen died. It is possible that the members of a fire company are better judges of a chief's qualifications than is city hall.

The same writer mentioned the problem of drinking in volunteer firehouses, which sometimes keep beer in a refrigerator or dispenser. Companies, of course, vary in the discipline they enforce, but those which have beer in the firehouse often lock it up until after a fire. There is no question, however, that fire fighters who may be called out unexpectedly at any hour of day or night may have had some drinks in a bar or at home. A paid man may be forbidden alcohol during his working hours, but a volunteer is on call 24 hours a day. However, in a company with good discipline the officers will put a man on the sidelines if he has had too much to drink.

A volunteer fireman with over 30 years experience states that intoxicated householders are far more often a problem than drinking fire fighters. One of the two or three leading causes of fire is "careless smoking" which usually means a half-plastered man or woman dozing off in bed or on an upholstered chair and dropping a burning cigarette. The fire fighters sometimes carry out their corpses.

A smoldering upholstered chair can cause a $20,000 fire in an expensively-furnished home. Not all such fires, of course, are due to alcoholic befuddlement: there was one in a tree house where kids had been surreptitiously smoking in an old chair.

The tradition of self government in fire companies has some drawbacks, but it tends to make modern companies law abiding and responsible: they can screen prospective members and expel those who misbehave. In the brawling days of the past it made it possible for a company of rowdies to take in their cronies, but at present when most companies are made up of responsible citizens the system works well within limits.

One of these limits is ethnic. In areas with a high concentration of one ethnic group the fire company is likely to be almost exclusively Italian, Irish, or Polish, as the case may be. But whether one such group or WASPs predominated there has been until recently an almost total exclusion of blacks. Until the courts struck down such clauses, the bylaws of many companies specified white membership.

This discrimination goes far back. In 1818 when some Philadelphia blacks attempted to form the African Fire Association, white companies held a joint meeting which resolved:

> The formation of fire-engine and hose companies by persons of color will be productive of serious injury to the peace and safety of citizens in time of fire, and it is earnestly recommended to the citizens of Philadelphia to give them no support, aid or encouragement in the formation of their companies. . . .

The meeting appointed a committee to visit the authorities and request them to prohibit African companies from opening fireplugs.

Meanwhile, the blacks, foreseeing trouble to their race, met and adopted a resolution saying:

> A few young men of color had contemplated the establishment of a fire or hose association; and although the same may have emanated from a pure and laudable desire to be of effective service in assisting to arrest the progress of the destructive element, we cannot but thus publicly enter our protest against the proposed measure, which we conceive would be hostile to the happiness of people of color, and which as soon as known to us, we made every effort to suppress. Should it be carried into effect, we cannot but consider that it will be accompanied with unhappy consequences to us. Therefore we sincerely hope that supporters of the contemplated institution, and such as might wish to be concerned, will relinquish all ideas of the same.

Following this appeal the members of the African Fire Association met and passed a resolution of regret at the "erroneous construction put upon the undertaking," and desiring to vindicate themselves from unjust implications and to "assert the rectitude themselves from unjust implications and to "assert the rectitude of their intentions as they were influenced solely by a wish to make themselves useful" declared that they "did not expect dissatisfaction, or they would not have progressed so far." They resolved to dissolve the Association, and return the subscriptions.

The abject tone of the surrender tells much about the precarious status of Philadelphia blacks—a status to be brutally demonstrated by white mobs and arsonists in the following decades.

Until well into the twentieth century, the blackface "Darktown Fire Brigade" was a feature of firemen's parades. This caricature was enshrined in a series by Currier and Ives.

As late as August 1972 a court order requiring New Jersey fire companies to include blacks in their membership brought a threat from three companies in Neptune Township to stop operations. Police Chief Elwood Seeds said the move came because the members of the volunteer companies believed "that the state has usurped their constitutional rights [by] not allowing them to vote on their own members." Undoubtedly discriminatory practices in many companies never come before the courts.

However, in 1976, of 45 chiefs of volunteer and part-paid companies across the country who responded to a questionnaire, 21 listed their companies as racially integrated. To what extent this represented substantial or token integration is not clear. As might be expected, companies in areas with few blacks were not integrated. The highest proportion of racially integrated departments were in Pennsylvania and the South. Maryland, Virginia, North and South Carolina, Florida, and Texas all reported integrated departments. In fact, antebellum New Orleans fire companies enlisted blacks to help pump engines, though not, of course, as regular members.

The admission of women members to fire companies has faced similar opposition. To be sure, in the days of bucket lines women and children took part. An old New York fireman remembered Molly, a slave, who acted as a volunteer in Engine Company No. 11. Once during a snow storm in 1818 she helped to pull the engine and who ever after that described herself as a member of the company. And the remarks of another old-timer about women fire fighters have been quoted.

However, the inclusion of women as regular fire fighters in both paid and volunteer companies is a recent phenomenon, a part of the women's lib movement which has carried women into the armed services, the police, and into trades and professions formerly considered male preserves. Recent legislation and court decisions forbidding discrimination because of sex have accelerated the acceptance of women by fire companies, but an important cause has been the spread of dormitory suburbs and the tendency of 18- to 22-year-olds to go away to college. As has been mentioned, many communities tend to become depopulated of able-bodied young men between the daytime hours of 8 to 5.

Long before women were accepted as fire fighters many volunteer companies had women's auxiliaries whose functions included serving at social affairs, supplying coffee and sandwiches at a fire, and perhaps furnishing first aid. Often they served in the ambulance corps, either as drivers or medical technicians.

In at least one company membership in the auxiliary led to acceptance as fire fighters. In 1970 seven members of the Women's Auxiliary of the Princeton Junction Volunteer Fire Company No. 1 began to help drive the trucks and handle hose during fires, but they did not enter burning buildings.

By 1974 two women were accepted as full-fledged fire fighters and entered the Mercer County training program. The company changed its bylaws to call everyone a *fire fighter* instead of *fireman* and to replace *he* with *he/she.*

Captain Larry Boyer said that when women were first employed at fires, "We took quite a bit of flak from surrounding companies. You'd call them male chauvinists today. But the women more than proved themselves to be worthwhile."

The flak did not come only from male chauvinists. When Arlington, Virginia employed Judy Livers as the first paid woman fire fighter in the country, the wives of firemen marched in protest and called Chief Robert Groshon "a dirty old man." In paid companies there was the problem of separate sleeping facilities. In Del Mar, near San Diego, California, Sue Martens slept in a couch in the chief's office because the city could not afford to build separate facilities.

This was not a problem in volunteer companies, but the question of women's physical capability was bound to arise. However, women soon demonstrated that they could pass fire training courses. Carmen Polk, who had served as a volunteer when Micanopy, Florida, organized a volunteer department in 1974, became the city's only full-time fire fighter because she was the only person who qualified for the job. A former nurse's aide and hospital ward clerk, she took on full responsibility for keeping the Micanopy Department running during the day.

Chief Joe Zuccarello of the Princeton Junction Company pointed out that although he is 5 feet 7 inches tall and weighs 125 pounds he can handle a 97-pound length of hose, but "even I have to struggle with it." He stated a truth every fire fighter knows, "You can do a lot more in a fire situation than you can do normally. The old adrenalin really gets flowing."

A survey of 40 departments by the magazine *Fire Command* in December 1975 found that not counting women's auxiliaries there were more than 250 women fire fighters in rural, semi-urban, and private fire departments. Overall in those departments which answered, 4 percent of the fire fighters were women. In one department in Mississippi three out of four volunteers were women; whereas in the Baltimore City Fire Department they number only five of the 2,500 volunteers. The largely volunteer department of Butte Falls, Oregon and the Gasquet Fire Protection District in Northern California had women fire chiefs.

The Annual Report of the Michigan Firefighters Training Council for 1975 noted that to date 73 women have been certified after completing the 66-hour training course. Mrs. Margaret Olbrick, the first woman to complete the 21-week Mercer County [N. J.] Fire School finished with the excellent grade of 82 percent. Mike Johnson, a fire fighter and member of the Board of Directors of the Parker Fire Protection District of Colorado said that a woman's division was formed in 1973 but that two years later the division was so well trained that it was disbanded and no distinction is now made between the sexes. "Men and women are on an equal standing."

One problem has developed, that of protective clothing: men's gear did not fit the women. Both the Globe Manufacturing Company of Pittsfield, New Jersey, and Servus Rubber Company of Rock Island, Illinois made changes in their product lines to accommodate women. Servus found sales disappointing, but Globe, the largest manufacturer of protective outer wear, reported a substantial amount of business, chiefly from smaller volunteer departments where women staff the companies during the day.

However, the supervisor of the Maryland Training Academy, as reported by the Michigan Firefighters Training Council, found it necessary to warn that the wearing of earrings was a dangerous practice, especially for persons with pierced ears. The long dangling type could get entangled in the head harness and prevent proper sealing of the facepiece of breathing apparatus, or it could get snagged and tear the earlobe. Metal earrings could conduct heat and plastic ones could melt, causing a serious burn. (In fact some companies frown on bushy sideburns on men for similar reasons.)

It is significant that the ancient tradition of membership in a fire company passing from one generation of a family to another also applies to women. Mrs. Olbrick, referred to above, noted "my dad and my brother have been in the company for years now." In Princeton, Hannah Rodweller, the borough's first woman fire fighter, had two generations of firemen behind her. The grandfather of Carol Rambo, a fire fighter in Warminster, Pa., was a fire chief; her uncle is chief of her company; her husband and cousin are members.

As is the case with racial integration, the admission of women fire fighters is a slow process. Only two of about 46 volunteer fire companies responding to a questionnaire for this study reported women fire fighters. One chief emphasized *no* by four X's. Governmental pressure is perhaps more easily brought to bear on municipalities employing paid fire fighters. For instance, the New Jersey Division on Civil Rights is insisting on strict compliance with the laws against sex discrimination. Other states take similar action with regard to equal employment, but it is difficult to prove that a company which votes on its own membership is doing so on a racial or sex basis. Furthermore, a person discriminated against for paid employment might sue; someone denied membership in a volunteer company might not.

However, in Delaware County, Pennsylvania a suit begun in 1975 by Beverly Epright and Maryann Gats led to a court order three years later requiring the 400-man volunteer company of Parkside to knock out any rules or policies which barred an applicant because of sex. These 23-year-old medically trained women did not want to fight fires but to answer ambulance calls and have the right to vote in the company.

The Pennsylvania Justice Department is trying to convince the 2,500 volunteer companies in the state that they'd better let women be fire fighters—or else departments which discriminate against women will face the loss of state-funded low-interest loans for new equipment.

Like all societies with a veneration for tradition, the volunteer fire company shares the values and limitations of such groups: on the one hand cohesiveness, on the other exclusiveness. However, the tide is moving. Firemen will not long be able to hold out against liberated girl friends, sisters, and wives. The women have the advantage of need for their services. Dormitory suburbs and a high proportion of 18- to 22-year-olds in college mean that in many communities the young married women are the best pool of available volunteer fire fighters.

Exhausted fire fighters resting after battling a brush fire that burned 250 Santa Barbara, California, area homes in 1977. *(Wide World Photos)*

Gladwyne fire fighter with the decoy rescued from the Schuylkill River, January 10, 1977 (The Evening Bulletin, *Philadelphia)*

George Johnson, 84-year-old Chief of the Jamestown Rural Fire Department, North Dakota, 1978.

Fireman priest—Father Joe Krustenski, a Plantville, Connecticut priest is a volunteer fire fighter. While he prepared to say Mass one Sunday, the alarm rang. Father Krustenski left to man a deck gun in a battle against a two-hour warehouse blaze. *(Wide World Photos)*

The Volunteer Fire Department of Franklin, Michigan, population c. 3,500. A fairly typical small town department, but with an unusually handsome firehouse.

The American LaFrance CU-1 Century of the Ellington Volunteer Fire Department: The control panel shows the numerous gauges and valves that must be learned by the operator. In a volunteer company, a number of members must be trained in pump operation to insure that at least one for each engine will show up at a given time.

The funeral of 21-year-old William Caber of the Woodlyn Volunteer Fire Company, Delaware County, Pa., who died while fighting a house fire. Three hundred firemen from Delaware County and surrounding areas attended the funeral. Aerial ladders are symbolically raised *(The Evening Bulletin,* Philadelphia)

American LaFrance CUS-2 Century 85′ Snorkel

SCORE CARD

How EFFECTIVE ARE volunteer fire companies? That is much like asking how effective are police forces. In the first place there are no very meaningful statistics. In the book, *Local Fire Administration in Pennsylvania,* published by the Department of Internal Affairs, Elizabeth Smedley discussing the factors determining insurance rates, stated:

> Several factors are not considered in determining the classification of a municipality. For example *there is no way of giving credit for the efficient performance of a fire department as measured by low fire loss.* [Ital. mine] Fire losses to be an effective gauge of efficiency, would have to be determined over a long period of years and would have to be weighed against the total amount of burnable property in a municipality. Even so, there is such an element

of chance that the use of this factor would not make the ratings more accurate.

Among the seven factors considered water supply ranks first; the fire department second. Other factors are the presence or absence of a fire alarm system, police, building laws, hazards, and structural conditions.

As to water supply, the most casual observer will note that a big city has many more hydrants in a given area than does a sprawling suburban community or the average small town. Rural areas, of course, have none. Furthermore, many suburban and small-town fireplugs may look more effective than they are. Many are connected to relatively small water mains, and to make matters worse, several in a given area may draw from the same main. Thus, a second pumper hooking up to a hydrant a block or more away from one in use may cause the first one to draw a vacuum, as the phrase is. By contrast, large cities often have special large high pressure mains exclusively for fire hydrants.

Thus most paid departments attacking a large fire have the immense advantage of adequate water supply, an advantage denied to many, perhaps most volunteer companies. Water companies are often better at keeping up their rates than their pressure. It is sometimes necessary for a fire chief to call the water company and plead for enough pressure to supply a pumper.

A most embarrassing example of this happened recently when two suburban fire companies in the Philadelphia area held a drill in which they set fire to a row of abandoned houses, only to be unable to get water out of the hydrants. The fire knocked out electric power for several hours; 150 nearby residents had to be evacuated; and five neighboring fire companies were called to assist. It was only by securing water from a naval base in the area that the blaze could be controlled.

Suburban and exurban mansions on big estates with stables and garages are often a half mile or more from a hydrant. That means a long hose lay and two or three pumpers hooked up in relay. A swimming pool helps if it is not empty, frozen over, or inaccessible to a pumper.

Adequate water supply, a police force, and a fire alarm system helps to explain why a city like Reading, Pennsylvania, with a volunteer system, is in the same insurance classification as Philadelphia and Pittsburgh with paid departments; in fact Reading's rate on a comprehensive homeowners policy is lower than that in the other two cities.

There is less difference than is often supposed in the comparative times of response between paid and volunteer departments. Volunteer companies which employ paid drivers are on the road almost as soon as they get the doors open, especially in the daytime. Except for the hours between midnight and 7 A.M. there are usually at least two or three company members in the firehouse playing cards, darts, pool or Ping-Pong, watching television, or just shooting the breeze. A pool table and a color TV set are invaluable assets in a firehouse. For, as has been mentioned, the fire company is also a social club—a function not to be deprecated.

Even at night the response is prompt. Reporting on his experience as a volunteer fireman in Weston, Connecticut (pop. 6,700), James Daniel, a writer, told of a fire at 2 A.M.: "All over town salesmen, lawyers, carpenters, dentists, plumbers, shopkeepers, construction men, and engineers rolled out of bed and into their boots. Within three minutes some of us were barreling down the road on the big red and gold firetrucks. Others learned of the location of the fire by phone or shortwave radio and converged on the house in cars." Five minutes after the alarm the men had water on the fire, and nine minutes after arriving had the fire out."

Even allowing for what Mark Twain called embroidery,

that is a rather typical response time if there is a paid driver
on duty or if one lives within a door or two of the firehouse.
Rural companies or those in scattered communities are necessarily somewhat slower to respond.

The shortwave receivers tuned for fire calls have been a
great boon to the volunteer company. They can alert men in
cars, at work, or in homes out of earshot of the siren. A
survey in 1976 shows that they are widely used. In some
companies, especially in rural areas, fire fighters keep their
turnout gear at home, and, directed by radio, go straight to
the fire; more commonly they pick up their gear in the
firehouse and look at a blackboard notice of the address of
the fire. The widespread use of alerting units has greatly
improved the performance of rural and small-town companies. However, one problem in such companies is that
apparatus not regularly maintained or frequently run is not
always in good condition. Batteries run down, tires go soft,
oil seals in, the pumps dry out.

The greater effectiveness of part-paid companies—which
usually means those with two or three paid drivers—is revealed in a study of fire protection in 1966 in Montgomery
County, Maryland.

	LOSS PER $1,000 VALUATION
Volunteer	2.85
Mixed (part-paid)	0.78
Paid	1.05

It would seem that the combination of paid and volunteer
fire fighters is especially efficient; but as has been mentioned

earlier, other elements affect fire loss, not the least of which is pure chance. For instance, in Lower Merion Township, adjoining Philadelphia, an area served by seven volunteer companies with paid drivers, the Belmont Hills Fire Company had a loss of $141,975 in 1975 as compared with $29,400 the year before. Much of the 1975 loss was due to one fire—a gas explosion which demolished an expensive home. No fire company could have saved it: the house blew up before an alarm was turned in. On the other hand, the neighboring Penn Wynne Fire Company had a loss of $5,900 in 1975, but $103,025 in 1974.

As of 1957 the National Board of Fire Underwriters rating Pennsylvania cities found five with the fewest deficiency points—one of them being New Kensington with a population of 21,000 and five volunteer fire companies. As of 1976 the rates for a comprehensive homeowner's policy on a $30,000 masonry house were $210 in Philadelphia, $137 in Pittsburgh, and $107 in Reading.

Certainly volunteer and part-paid departments save their communities a vast sum of money. Questionnaires for the 1968-1969 National Fire Service Directory of the National Fire Protection Association were sent to all departments protecting communities of 12,000 or more in the United States and Canada. These revealed that the average cost per person protected by 30 volunteer fire departments was $2.25 a year; for 14 departments with one to five paid men, the cost was $2.83; and for large departments with 100 or more paid fire fighters it was $13.69. Thus volunteer or call departments cost on the average one sixth of the amount per person protected by paid departments.

As of 1968 it was estimated that the volunteers were rendering a public service worth at least 5 billion dollars. With the subsequent inflation the amount might be nearly twice that. And as the report concluded, "Certainly, the interest and enthusiasm of the typical volunteer is something money cannot buy."

Granting that volunteer companies, especially those with paid drivers, can function efficiently under normal conditions, what about their ability to handle major disasters? A few examples will help to answer this question. A few years ago the Jamestown, North Dakota Rural Fire Department with a membership of 17 was faced with a fire in a grain elevator holding about a million bushels of wheat. Like a coal mine fire this one could not be extinguished by simply pouring water on it, and furthermore, the loss would have been tremendous.

It took just two days short of three months to extinguish the fire. All told, the little fire company made ten runs to keep the smoke down while the grain was being removed. On the last day the company was on the job from 9 A.M. to 9:30 that night.

Although small-town and suburban companies are not primarily set up to handle industrial fires, they increasingly have to cope with them. On the morning of July 31, 1956 a demolition workman with an acetylene torch cut into an oil pipeline in the abandoned Autocar truck factory in Ardmore, Pennsylvania. The demolition men tried to fight the resulting fire, but it spread rapidly through the old oil-soaked five-story brick building. A barber across the street turned in the alarm at 10:56 A.M..

To the east of the plant, separated by an alley, was a business district about two blocks long consisting of small stores, restaurants, automobile show rooms, and the Lower Merion Township office building and police headquarters.

Ardmore is a community of about 9,400 population, situated on the main line of the Penn Central Railroad, nine miles from center city Philadelphia. It is the administrative center of Lower Merion Township, a wealthy suburb which stretches from Philadelphia westward to Villanova, with the Borough of Narberth in the middle. The area in 1956 had about 60,000 inhabitants and was served by seven volunteer fire companies.

One Autocar factory, established about 1900, had spread over a block-long site along the Lancaster Pike, which runs through the center of Ardmore. Immediately behind the plant was the railroad, which prevented fire trucks from access to the area except at one corner of the buildings. In addition to the main building there were several frame structures dating from the early days of the company. The location in the middle of a largely residential suburban community had long been unsuitable for the manufacture of motor trucks, and the company had recently removed to Exton.

Long before the fire the nearby volunteer companies had pre-planned their strategy in the event of an emergency at the plant. Thus when the call came, each company took up a previously-assigned position. In view of the size of the conflagration four other volunteer companies were called to assist, and a truck with a deluge gun was sent by the Philadelphia Fire Department. As is the custom when several companies are in action, neighboring companies moved in to stand by in empty firehouses.

For a time it seemed that the business district of Ardmore might go, but with some luck in wind direction the firemen confined the blaze to the factory. The fire halted traffic on the railroad, but hose lines from the other side had to be detoured through underpasses at some distance from the plant. During the day-long battle 200 firemen pumped 5 million gallons of water on the blaze.

Yet despite the inferno of a huge oil-soaked building and falling brick walls there were only minor injuries: a face burn, a bruise from a nozzle, a cut hand, and a nail in the foot. Pre-planning and skilled direction by Fire Marshal James Mullen and the company chiefs had saved the center of Ardmore and prevented serious casualties.

In May 1969 the Eaton Rapids, Michigan Fire Department was called to a fire in a large woolen mill. On duty was Chief Charles W. Oliver, one of the three paid men in the department. After sounding the siren and alerting the volunteers on their home receivers he started for the mill which he saw was fully involved. Within four minutes after the initial alarm a call was sent to the departments in six surrounding towns.

The men found water pressure enough for only one 2½-inch line and the sprinkler system in the mill had ruptured. The Delta Fire Company, using a different hydrant, used its one line to protect men covered by a wet tarpaulin so they could get to the shut-off valve for the sprinkler system. The assisting companies hooked up to distant hydrants to supply the Eaton Rapids Company, but they also extinguished roof fires in a barn and three houses plus an unrelated house fire caused by faulty wiring.

Although two sections of wall fell, the men got out of the way without injuries. In all seven companies only two men were overcome by smoke and heat and the chief cut his wrist on broken glass. The alarm had come in at 12:37 P.M.; by 3:30 P.M. the fire was under control.

In Locke, New York a tank truck rolled over pouring burning gasoline down the street into a bank, a garage, a diner, an antique auction building, and two houses. Because the water system was inadequate, the pumpers had to work from tank trucks and from static water sources such as ponds. That meant hooking the engines in relays. Altogether 20 departments sent 22 pumpers, 15 tankers, two ambulances, and two emergency trucks. To fill the empty firehouses apparatus from five counties moved in. A force of about 600 men was involved. Only volunteer companies could have marshaled so much equipment and manpower.

For instance, one study found that the average number of paid firemen in the larger departments was 1.62 men per 1,000 population; for fully volunteer or call departments it was 4.06 men per 1,000, or 2.5 time as many. Of course, volunteer departments need a larger enrollment than do paid departments because many volunteers are not available at certain times, especially during the day. However, it must be remembered that only a third, or at most a half, of a paid company is on duty at any given time.

When a fire requires many fire fighters such as one in a large tract of woodland or brush, or in a lumberyard or industrial plant, volunteers tend to come in from all over the place to join with those who reached the scene first. For a fire after 5 P.M. or on weekends volunteer companies can marshal most of their personnel.

This was the case when on the night of January 27, 1971, with a high wind and a temperature of 10 degrees Fahrenheit (chill factor minus 27°), gas explosions in two adjoining houses in West Conshohocken, Pennsylvania touched off a fire fed by three gas mains under the street, the largest being a 20-inch feeder line. The little borough with a population of 2,162 has an excellent volunteer company—for some unknown reason named the George Clay Fire Company—which responded immediately. By that time all electrical power was out and a whole block of houses was on fire. Flames were shooting 50 feet in the air.

As the men on the first truck started to fight the fire another house exploded, killing a George Clay fireman. Men on the second truck escaped because the power failure delayed the opening of the firehouse door. Altogether 18 volunteer fire companies responded with 250 men. Until the Philadelphia Electric Company finally got the gas shut off three hours later, the firemen could do little about the blazing houses. A huge gas fire cannot be put out with water and even a small gas fire must not be doused for the escaping gas will promptly explode.

What the firemen were able to do despite hose lines and hydrants which froze was to protect a lumberyard at the end of the block and various nearby industrial structures. As soon as the gas was turned off they doused the burning houses. One company returning home found a small trash fire but discovered that everything on the engine was frozen up. They used a bucket to dip water from a creek.

The disaster caused the death of a fireman and two children, and injuries to 26 firemen and 21 other persons. The cooperation among the 18 fire companies could not have been better.

As with the Autocar fire, good volunteer departments pre-plan their strategy in case of a fire in an apartment building, school, or industrial complex. Thus at 9:50 P.M. May 3, 1974 in West Hamstead, New York when a fire started in the millwork shop of a lumberyard, the Lakeview Fire Company was prepared. Adjoining the lumberyard was a gas station with a repair shop and a contractor's storage yard containing roofing, lumber, and nearby the contractor's trucks. Adjacent to the storage shed was the Long Island Railway surrounded by fencing and with a 69,000 volt transmission line.

Chief Richard Sena immediately requested mutual aid from the West Hamstead, Malvern, and Rockville Center Fire Departments. In accord with company strategy the companies laid seven 2½-inch lines, two 1½-inch lines, set up two deluge guns and three ladder pipes. By 1 A.M. the fire was under control. The volunteers had saved the lumberyard and the slightly-scorched gas station, and prevented damage to other property.

Probably the most dreaded fires are those in oil refineries and storage tanks. Many of these are in areas served by volunteer companies. Professional fire departments do not always handle oil and gasoline fires well: in fact a film used for training purposes in volunteer companies shows firemen in a large, paid department attacking a fire in a tank truck at a loading platform in a manner which led to the deaths of several men. Using 2½-inch hose lines they drove the fire back under the storage tanks, the ends of which faced the firemen. Now any fire fighter who has taken a training course has been warned that if a fuel tank, whether stationary or on a truck, explodes, the ends will blow out. That is what happened in Kansas City: a tank took off like a projectile, landing among the firemen and engulfing them in flaming gasoline.

Following a Gulf Oil Refinery fire in 1975 which killed eight Philadelphia firemen, a reporter visited volunteer companies in nearby New Jersey areas which have refineries. Asked how he would feel about going to a fire in the Mobil refinery like that in Philadelphia, Bill Suiter, the assistant chief of the Gibbstown Volunteer Fire Company said he'd be scared but he'd go. Over the years the Gibbstown Company has gone into Mobil to help the firm's own department. Back in 1951 Gibbstown and two neighboring volunteer companies fought a six-hour blaze there that involved exploding oil drums. A few miles away in 1970 the Vega Volunteer Fire Company fought a fire in the Texaco refinery for 32 hours. .

Oil companies with a long "the-public-be-damned" tradition do not always make it easy for firemen. Whether or not the Philadelphia Fire Department handled the Gulf fire well, much of the responsibility for the fatalities was due to the failure of employees to turn off strategic valves. The Vega

firemen criticize the Texaco Company for lack of cooperation. For one thing it will not permit drills on the refinery
grounds.

On the other hand, the Mobil Company permits the Gibbstown crews to come in for regular drills supervised by company experts.

As is customary elsewhere, companies from all over
Gloucester County would be called under mutual assistance
agreements in the event of a bad refinery fire. In the largest
fire in the county's history, that in a four-story block-long
factory building on March 9, 1970, more than 55 pieces of
equipment and 400 men were marshaled from 32 Gloucester
and Camden County fire companies. As the reporter discovered, "Though all of Gloucester County's fire companies
are volunteers, they can mount an impressive array of equipment and manpower."

One of the worst fires ever faced by volunteer firemen was
the *Corinthos* disaster in Marcus Hook, Pennsylvania at
12:30 A.M. on Friday, January 13, 1971 as the Liberian
tanker *Corinthos* docked at the British Petroleum refinery
pier and, loaded with approximately 300,000 gallons of
crude oil, was rammed by the *Edgar M. Queeny,* an American chemical carrier. Within ten minutes there were no less
than five major explosions which rocked the town and could
be felt 20 miles away.

Marcus Hook is a town of 3,000 at the southeastern tip of
Pennsylvania and bordering on the state of Delaware. Its
volunteer fire department consists of the Marcus Hook Fire
Company and the Viscose Fire Company, which operate as
separate administrative bodies but upon an alarm, work as a
joint department. In both companies the field officers rise
through the ranks holding office for four years. The head
man of each company then assumes the office of Assistant

Borough Chief for two years and finally Borough Chief for two years, the posts alternating between the companies. These are nominally paid offices.

At the time of the disaster the Fire Department Chief was Robert Sides, a member of the Viscose Company for 12 years and who had completed all fire science courses at the college level and had attended local and state fire schools throughout his career. When not engaged in fire department duties he is employed as an Advertising Representative for the *Delaware Daily Times* in Chester.

Chief Sides was on the first unit at the scene where the men were faced with flames reaching 400 to 500 feet in the air and singeing their skins. The first alarm brought out approximately 60 men from the two companies, and a second alarm brought companies from Linwood and Boothwyn, which also dispatched ambulances. Additional ambulances were at once requested from eight other companies to take exposed seamen and dockworkers to area hospitals.

The initial concern of Chief Sides was to stop the fire from advancing into the refinery area. The Marcus Hook companies have had much experience with industrial fires. A Sun Oil Company, which occupies about 50 percent of the town's area, has had numerous fires, calling out the borough companies an average of once a year. The worst was in 1946 when 11 persons, including several volunteer firemen, were killed. The other refinery, British Petroleum (BP) had a policy of handling emergencies themselves until the *Corinthos* fire. In addition to the two refineries there are numerous industrial plants including two large ones belonging to Allied Chemical; several facilities of the FMC Corporation which produce cellophane and assorted textiles, and a Congoleum plant manufacturing vinyl floor coverings. Although Sun Oil and FMC have fire brigades, all these companies except BP call on the Marcus Hook Department.

Marcus Hook's second pumper and the Viscose pumper fed hand lines and deluge guns used to cool down exposed oil pipes and the Naptha/Solvent area. During the first 30 minutes of operations there were continual explosions on the tanker and fireballs soared 500 feet into the air. Embers, rivets and debris from the ship were thrown hundreds of feet into the refinery. Despite the possibility of being engulfed, the firemen stuck to their posts, eventually pushing the fire back along the pier.

Burning fuel from the tanker flowed on the surface of the river, endangering the bulkheads and threatening the refinery and an oil storage warehouse. With this in mind Chief Sides called for additional equipment from nearby communities. Two companies put out a fire caused by the burning oil which reached a destroyer with 100,000 gallons of oil on board.

About a half hour after the initial alarm County Fire Marshal George Lewis arrived on the scene and with Chief Sides set up a relief system of four- and eight-hour shifts for round-the-clock operation whereby a company or department could volunteer for duty and report to a designated location. All in all some 50 volunteer fire companies were involved.

Food service to the firemen was provided by the Salvation Army and the Moyamensing Juniors, a group of young men who operate out of the Chester Fire Department, but the largest volunteer auxiliary service was provided by women from the local fire company auxiliaries. Joined by some men they set up a round-the-clock food service, supplied changes of clothing, and cots for sleeping. The women from the two Marcus Hook companies worked in shifts to furnish complete meals for fire fighters from all the participating companies.

By daylight on Friday the Marcus Hook Department and

neighboring companies had confined the fire to the dock and the area immediately adjacent to the burning ship. Now the problem was the burning oil. The only solution for this is foam—in this case vast quantities of the expensive stuff. A large supply was made available from three sources: BP's own stockpile, Sun Oil, and most importantly, trailer loads from National Foam's West Chester plant.

With advice from experts of the foam company the fire companies got the foaming operation under way about 6 P.M. One of the two fireboats from Philadelphia was enlisted in the foam operation. Almost 100 firemen formed a human chain to hand 5-gallon containers to the end of a nearby pier where Coast Guard boats picked them up to take them to the fireboat. Altogether about a quarter of a million dollars worth of foam was used.

Nevertheless, at about the time the foam operations were begun at 9 P.M., the fire erupted from the bow of the ship, a prelude to the release of a huge amount of burning oil. The 30 volunteer firemen on the dock at the time were ringed with fire and had to jump onto the deck of the fireboat.

By Saturday evening the fire on board the *Corinthos* was largely confined to the bow and stern sections. By daylight Sunday little fire was showing, and units from Primos, Linwood, and eventually Marcus Hook were released. Coast Guard Chief Soper, who had specialized in fire fighting during 34 years of military service, said: "It equals anything I have seen, and that includes Pearl Harbor and Guadalcanal." The volunteer firemen had fought the blaze for 60 hours.

In this era when flammable and toxic materials are constantly in transit, small companies outside of industrial areas may be faced with blazes and explosions of hazardous materials. On October 18, 1976 a Chessie freight train derailed outside of Clifford, Michigan, population 500. Moments later there was a terrific explosion followed by flames visible 50 miles away.

Clifford Fire Chief Mel Sarles got the call at 5 A.M.: "We got us a train wreck!" His 16-man company had only one 500-gallon pumper, a 1,000-gallon tanker, and a brush truck. At the scene they found two tank cars ablaze, one containing butadiene, the other acrylonitrile—both highly flammable. Butadiene, used in plastics, evaporates so rapidly that it can cause frostbite on contact or destroy the respiratory system if inhaled. Acrylonitrile contains cyanide.

Sarles called for help from five neighboring companies, and contacted the local office of the Dow Chemical Company. He was told not to throw water on the blaze. Already many of the men had begun to feel the effects of the gasses. Said fire fighter John Young, "It was like a sudden epidemic of sinus trouble."

With the spread of noxious clouds of gas, the police and fire officials evacuated some 2,500 people from Clifford and surrounding communities.

Meanwhile, the volunteers laid hose lines from the top of an adjacent hill. When the Dow experts decided that the acrylonitrile had burned itself out, the firemen moved back in. Because there was not much the other companies could do, Chief Sarles sent them home, and his company settled down for a long vigil. During the first night it was cold, but the flames kept shooting up 50 to 100 feet in the air. Sometimes the wind would shift, causing the smoke to come down on the men.

By the second night the fire fighters were nearly exhausted, and no one was certain of the possible effects of continual exposure to butadiene fumes. To keep awake the men told jokes and kept punching each other. Other companies brought fuel for the trucks, and townspeople furnished coffee and sandwiches, but the men's chief desire was to go home. Finally, 76 hours after the start of the fire, the Clifford volunteers were able to douse the tanker.

However, their problems were not over: one man became violently ill from smoke inhalation and had to be hospitalized. As a precaution the others were also taken to the hospital, where it was found that their bodies had been seriously deprived of oxygen. They were kept 36 hours and fed oxygen mechanically. Chief Sarles was told to stay in bed for a week.

Like the firemen of a century before, he said: "I didn't regret one moment of the effort. I've been a fire fighter since I was 15 years old. I hope I'll be a fire fighter until I'm 60— I'll always do everything I can to help the people here."

Volunteer companies operating, as they often do, in areas with limited or nonexistent police or rescue services, are called upon for all sorts of duties. Late in October 1948 a temperature inversion caused a lethal fog loaded with industrial gasses to close down on Donora, Pennsylvania, a borough of 12,300 population and a concentration of heavy industry. Before rain dissipated the fog, thousands were made ill and 11 died.

At the start, a parade of the volunteer fire company fizzed because it was almost invisible. When it got back to the firehouse Chief Volk, one of two full-time men, got a call for the inhalator for a man who was choking. As he later said, "We're not supposed to go running around treating the sick. But what the hell, you can't let a man die!" For the rest of the night Volk went about administering oxygen to people about to die. There was so little oxygen in the air that everytime he took his foot off the accelerator his engine would die.

When he reached a victim, he would throw a sheet or blanket over the patient, stick a cylinder of oxygen inside and crack the valve. "By God, that rallied them!" Volk didn't take any himself; instead he drank a little whiskey to ease his throat each time he got back. At the start the fire company had about 800 cubic feet of oxygen; Volk ended up by borrowing from McKeesport, Monessen, Monongahala, and Charleroi. He was laid up for a week afterward, but he had saved a lot of lives.

Volunteer companies in communities bordering on lakes or rivers have boats for rescue operations.* These are usually designed only for fishing out floundering canoeists, taking people from flooded homes, and recovering drowning victims. However, volunteer companies, relatively free from bureaucratic bumbling are ingenious at devising means for meeting special conditions. Thus the 400-man volunteer American Hose and Ladder Company No. 2 of Bristol, Pennsylvania has an unusual marine unit—an actual fireboat suitable for use in shallow waterways or flooded streets. Only 16 feet long, it can be carried on a trailer, but is powerful enough to pump 3,125 gallons a minute. Only three such units are known to exist in the nation. During the flood caused by hurricane Diana in August 1955 the Winsted, Connecticut Volunteer Fire Company No. 3 apparently did not own a boat, but eventually got hold of one. In the meantime fireman Scott Weed, policeman Farris Resha, and other firemen had combed the tenement blocks, getting people out of bed and onto high ground.

By midnight the overflowing river had begun to pour down the street at 20 miles an hour. Weed and the firemen got an extension ladder across rooftops to a house opposite one where people were trapped. They crawled across the horizontal ladder and threw a rope to the far side. Using the

---

* During a rescue attempt in 1975 the Gladwyne Fire Company's boat swamped in the Schuylkill River to the embarrassment of the crew. The Company raffled it off and bought a larger one. Then, one cold January Sunday morning, they were called to rescue an ice-bound duck. Breaking through river ice for a half mile they found a Canadian goose—a decoy. As one man said, "It's just one of the crazy things we do." Two years before they had gone to Philadelphia to rescue oil-soaked ducks. The news item brought letters of praise from as far away as Texas, Arizona, and Hawaii.

rope to pull a boat back and forth they rescued several people, including a 16-month-old baby and an elderly couple. On the last trip a young Italian woman panicked and capsized the boat. She fought off the man who tried to save her and she drowned. The firemen and Resha rescued two of the boatmen with ropes, and two others managed to catch hold of debris along shore.

When the rain stopped in the afternoon, all the rescuers went to Resha's family restaurant for their first food and drink in 18 hours. After the flood Weed worked around town pumping out cellars.

One of the most common problems faced by suburban fire companies is the car crash in which people are trapped. There are various tools to cut through or pry open automobile doors. In 1974 the Orangeburg [New York] Volunteer Fire Department bought a Hurst tool with five tons of prying and cutting power. During the next 15 months they got seven rescue calls and removed five victims alive. Two were dead when the firemen arrived. As often happens the dead were a ghastly spectacle. One especially difficult problem was faced by the Orangeburg fire and police departments when they responded to a call for a young woman trapped in a car. Within four minutes they reached the scene where they found the victim unconscious, bleeding, and looped in a U around a utility pole. A high tension wire attached to the pole threatened to fall. They administered oxygen, and using their Hurst "Jaws of Life" tool, they got the woman out alive.

It is worth nothing that 34 Orangeburg fire fighters have completed a standard Red Cross first aid and personnel safety course.

Rural areas are unlikely to have specialized equipment such as Jaws of Life tools, but the CB radio has become common. Its usefulness was demonstrated during a prairie fire which swept over 50,000 acres in Osage County,

Oklahoma on February 17, 1976. The first report was called in by a rancher to Mary Hazelbaker, "Grandma Base." The wife of a retired soil scientist who is fire chief in the small town of Fairfax, she keeps her CB base station on all day.

She immediately relayed the rancher's message to the Fairfax police captain who called out the volunteer fire department. In nearby Pawkuska, W. T. Wade, civil defense director there picked up the call. On the CB band he started calling ranchers, some of whom were out in their fields. They'd reply, "I'll be right over when I round up my boys."

One of the deadliest of natural disasters, a prairie fire can sweep over hundreds of square miles, jump over four-lane highways or a half-acre of ground. As Richard Elliot, chief of the volunteer department in nearby Skidler, said, "We had a 35-40 miles per hour south wind that day, and as it hit a pasture, it seemed like the whole 160 acres exploded at once."

On the CB radio Mary Hazelbaker was busy rounding up civilian help. Oil companies and ranchers sent tank trucks and trailers with water. Drivers of cars and trucks cruised the roads sending reports of trapped herds of cattle.

After 12 hectic hours the volunteer firemen and civilian helpers had the fire out. In addition to the 50,000 blackened acres, several houses and barns had been destroyed, plus an unknown number of cattle. Two men were burned seriously enough to require hospitalization. The county farm agent estimated the damage at $5 million. Without CB radio the disaster would have been much worse.

Volunteer fire companies near cities are sometimes asked to aid paid urban departments. On January 28, 1977 Buffalo, New York, which had already experienced 13 feet of snow, was struck by a blizzard. Fire rescue calls poured in: 50 for fire; 98 for the rescue squad. Apparatus broke down or could not get through streets clogged with abandoned cars; engines and trucks had to run with skeleton crews of four men. Calls for aid went out to surrounding communities which have 7,500 active volunteers. When some of these companies could not get through drifts as high as 25 feet, they mobilized toboggans, dog sleds, and dozens of snowmobiles.

Thousands of stranded motorists were housed in fire halls. Volunteer companies in Cleveland Hill sheltered 100; Cheektown, 500; and Bowmansville, 550. The Lake Shore Ladies' Auxiliary logged 380 hours caring for the refugees.

Because a number of volunteer companies had mini-pumpers (300 to 750 gallons a minute capacity) which could get through where larger apparatus could not, 18 of these got to Buffalo with 100 fire fighters. Volunteers came from as far away as Suffolk County on Long Island, 500 miles away. Six of these rigs with their crews were airlifted to Niagara Falls Airport where state troopers escorted them to Buffalo. Later, Suffolk volunteers charted a plane and flew in relief crews. During the emergency the mini-pumpers answered 417 alarms. Some of the volunteers stayed in the city for two weeks.

To paraphrase Dennis Smith's *Report from Engine Company 82*: if you call a plumber, he may or may not appear; if you use a pay phone, you may or may not get a dial tone; if you use an expressway, you may or may not reach your destination; but the one sure thing is that if you report a fire or an accident, the volunteer fire fighters will come.

In 1977 the firemen of Carteret County, North Carolina were holding a fund-raising all-male mock wedding when the alarm sounded. After a moment's hesitation the men in wigs, dresses, and make-up jumped on the truck and took off for the fire.

XI

THE HEIRS OF FRANKLIN

Children picking up our bones
Will never know that these were once
As quick as foxes on the hill; . . . .

WALLACE STEVENS,
A Postcard from the Volcano.

THE VOLUNTEER FIRE company is one of the few conservators of tradition in a nation notorious for its disregard of tradition. The Articles of Franklin's Union Fire Company 1736 provided for a meeting the last Monday of each month; those of the Hibernia Fire Company 1752 called for a meeting the first Monday of each month. The monthly meeting, usually on the first Monday, has become traditional in Pennsylvania and New Jersey. So too it is the meeting date selected by the recent company of Hobe Sound, Florida, and in other places across the country.

Meetings are conducted as they were 240 years ago, with roll call, treasurer's report, motions made and voted on democratically, and the appointment of committees. The old minute books frequently record the appointment of commit-

183

tees "to enquire into the state of the engine," and the later reports of such committees. Today a company will appoint an apparatus committee with the same function.

As the 1975 report of the Connecticut committee for the study of fire protection in the state pointed out: "Firefighting machines have become infinitely more powerful and sophisticated. . . . Yet in many ways things have not changed. Most of our muncipalities still have volunteer fire departments." A report of The National Commission on Fire Prevention advocating more attention to fire prevention, noted that the techniques of fire suppression have not changed greatly in two centuries.

After all, the first duty of a fire company is to get people out of a burning building, the second is to pour water on the fire. No doubt the fireman holding a nozzle in 1776 shouted the same call heard today: "Water! Water!"

The fire axe on the 1799 ladder truck in the museum of the Insurance Company of North America is identical with those in use today. And it was used in the same way: a hot wall must be opened or a roof puffing smoke at the eaves must be chopped through or the house will explode. Even a modern tool like an electric roof saw is used for the same purpose.

The design of most helmets in use today goes back more than 150 years; in 1820 the graceful sweep to the rear was introduced, and in the 1830s the famous New York maker, Henry T. Gratacap, added the raised frontpiece. (A recent type is a combination of traditional and motorcycle headgear.) The turnout gear worn by Nathaniel Currier, himself a fireman, in the 1858 Currier and Ives print *Always Ready:* helmet, boots, and waterproof coat is remarkably like that in use today. Currier would not look out of place at a fire in 1978.

The dalmatian or firehouse dog came in with horse-drawn

apparatus. Sometimes known as a carriage or coach dog, his function was to run behind the horses to keep off other dogs. Today dalmatians ride motorized apparatus. The Gladwyne Fire Company had one who used to jump off the engine and get in a fight with the house dog. Thus, before the men could fight a fire they had to break up a dog fight. When Sparky took to biting neighborhood kids, the company gave him away. A fire company heavily dependent on contributions cannot afford to alienate the neighbors.

The efforts of eighteenth-century companies and those after them to obtain apparatus better than that of their rivals has been recorded. That is still the ambition of companies today. It can sometimes go to excessive lengths, as in the community which voted down a school bond issue but passed one to buy a snorkel although no building in the town was over three stories. The firemen wanted to outshine rival companies in parades.*

In the past the volunteer fireman was alerted by the ringing of a church bell; later by one in the firehouse tower; and still later by a locomotive tire struck by a sledgehammer. The modern volunteer often has an electronic device in his home which gives him the location of the fire, and the firehouse siren or bull horn can be heard for miles. In any case the response is the same as two centuries ago. If the fireman is at work, he drops whatever he is doing, sometimes with the loss of pay, and answers the alarm. If he is in bed, he probably has arranged shoes, pants, or coverall for quick donning. Before the motorcar men ran to the fire or engine house; some still do, but nowadays a fire fighter normally has car keys ready on a bureau or bedside table.

Customs about the whereabouts of turnout gear vary:

---

* It is not only fire companies which engage in such rivalry. In 1899 the University of Pennsylvania changed the date of its founding from 1749 to 1740 so that the faculty could march ahead of Princeton in academic processions.

some companies allow fire fighters to keep it at home; more commonly it is on racks in the firehouse with the member's name or number above it. Geography has something to do with this: where fire fighters are widely scattered as in semi-rural districts, they probably keep coats, helmets, and boots at home; in more built-up areas they usually go to the fire-house to pick up their outfits, and ride or follow the engines. In either situation the gear is not fully interchangeable: a man who takes a 7½ helmet cannot wear a 6½ one; some-one who wears size 12 boots can't get into size 8; a woman needs a smaller outfit than a man. Therefore, turnout gear must be assigned and kept inviolate for a particular person.

As in the past most companies, in addition to turnout gear for fire fighting, also have some kind of dress uniforms for parades, funerals, and public functions. Depending on the prosperity of the company or the members dress uniforms vary from a white shirt, blue serge pants or skirt, and with some sort of insignia all the way to fully-tailored outfits similar to one worn by an admiral. In small communities the fire fighters usually buy these for themselves, but sometimes the company pays all or part of the cost. As the Goshen Fire Company budget shows, the members there pay back the cost of their dress uniforms.

In the past, firemen often spent large sums on presenta-tion helmets and parade trumpets for their officers. As noted, the Sacramento, California company presented their foreman with a $1,350 helmet made by the famous Henry T. Gratacap and encrusted with precious stones. The jeweler's bill alone was $840—the most expensive helmet on record.

The need for adequate fire fighting gear is obvious, but city sophisticates may think of dress uniforms as a silly affectation. That misses one of the most important features of a volunteer fire company—morale. It is just possible that part of George Washington's difficulty in keeping his army together was that the men were outfitted in such a haphazard fashion. A person in a handsome uniform carries himself or herself in a special manner; he or she tries to march and behave in the same seemly fashion as other marchers. For at least 175 years volunteer firemen have indulged in dress uniforms for special occasions. Such a uniform gives a gas station grease-monkey or a grocery clerk a sense of importance—as for that matter it does an admiral.

But whether a fire fighter is in a dress uniform or turnout gear he or she is a person apart, someone with a certain amount of authority. People will move off a sidewalk or out of a yard on orders from a uniformed person; wide-eyed little kids will ask respectful questions. For youngsters the fire fighter in turnout gear is automatically a hero.

After all is said, there is a strange mystique about being a volunteer fire fighter. He is a kind of folk hero. It is no accident that *The Life of a Fireman* series of Currier and Ives has been popular for over a century and still appears on calendars. And, of course, this series is only a small part of a vast iconography devoted to fires and fire fighting. Like the locomotive, the fire engine has been a favorite toy. But whereas the locomotive is fading away, and has certainly lost its glamour, the fire engine grows ever more handsome and ubiquitous, and the fire fighters who operate and ride the beautiful machines take on added stature like a policeman on a fine horse. One testimony to the attraction of fire engines and fire fighters is the difficulty of keeping kids out of the firehouse.

Today when a new $80,000 or $185,000 piece of equipment is obtained, it is usually housed with some traditional ceremony. In eastern Pennsylvania it is pushed into the firehouse three times as were the old hand engines. (Because of the weight of modern apparatus the driver uses the motor; the men pushing have a function similar to honorary pallbearers.) In other places the piece is "baptized"—that is, ceremoniously washed. Almost everywhere there are speeches and a meal. It is customary to sell glasses or beer mugs decorated with the company seal.

There is no similar ceremony for the acquisition of a new garbage truck, police van, gubernatorial or presidential limousine.

Firemen are enthusiastic about preserving the past. That too is a tradition: firehouses of a century ago often contained a small museum of ancient apparatus, fire buckets, hats, and other artifacts.

Firemen devoted countless hours to the making of miniature copies of hand pumpers, hook-and-ladder trucks, and steamers. The scores of such models owned by the Insurance Company of North America and the Home Insurance Company are testimony to the popularity of this hobby. Today the meeting room in a brand new firehouse is likely to have reproductions of the Currier and Ives series *The Life of a Fireman;* there will be pictures of old engines, and probably a cabinet containing old helmets, axes, trumpets, and fire buckets. There are likely to be toy fire engines, both antique and reproductions.

Companies often preserve their original or early pieces of apparatus. The Madison, Nebraska Company has a hand pump brought to the town in 1880, and the Nebraska City Fire Company still owns a steamer bought in 1884.

As might be expected, New England fire companies are especially rich in antique apparatus. And the members of a company will spend hundreds of hours restoring ancient pieces. The volunteer company of Townsend, Massachusetts has two old hand tubs; that of Greenwich, Rhode Island has an 1844 Button engine; the Leominster, Massachusetts company has an 1846 Hunneman; the Westminster company an 1849 Hunneman; the F.D. Hoxie Fire Department of Mystic, Connecticut has an 1883 Button, one of the last built.

These are only a few of the old hand tubs in New England. In 1860 the New England States Veteran's Firemen's

League was formed "to encourage and perpetuate the oldest sport in the Country," that is the competition at firemen's musters. These began in the early 1880s and continue to this day. At first the competition was to see which company could throw the highest stream, but because this was difficult to measure, it was decided to mark off the ground to measure the longest horizontal stream. Although musters featuring antique apparatus probably originated in New England, they have spread across the country. One in Valhalla, New York in 1977 drew 194 pieces of antique apparatus and hundreds of spectators from seven states. Another in Eureka, California attracted teams from all over the West.

In 1968 A Hunneman engine belonging to Swampscott set a record of 239 feet 11¾ inches. The 1882 Button engine belonging to Gardner, Massachusetts won a championship for seven consecutive years (1967-1973) with one record of 251 feet 11 inches.

This is much better than any of 17 companies did at an 1854 muster in Springfield, Massachusetts when the longest stream was 165 feet. At Hampden Park in 1870 hand engines were pitted against steamers with records of 171 feet and 272 feet respectively.

The explanation for the better performance of hand engines at recent musters may be due to hose with less friction and to the skill with which the old engines have been restored.

The Friendship Veterans Fire Engine Company of Alexandria, Virginia has got back from Baltimore and restored the 1775 engine presented by George Washington. They also have a hand pump engine made in 1849 and used at the 1851 fire in the national Capitol.

The restoration of a steamer is a more complex operation than the rebuilding of a hand engine.

The Townsend Company retrieved its 1908 Amoskeag steamer from a museum and arranged with the Fitchburg firemen to have it scraped, painted, and the boiler retubed. Finding that the old machine still would not pump, they undertook the long job of refitting 32 poppet valves and finding other parts. Three years after acquiring the steamer they had it pumping like new.

In 1971 the Wayne, New Jersey Fire Company, founded in 1918, acquired an 1899 steamer with an old Christie motorized front drive. For three years the firemen worked at restoration, getting the boiler retubed and going to a shop in Alabama for hard rubber tires. Twenty nine hundred parts were sent to be nickel plated. To pay for the restoration the firemen raised $18,500, a fifth of it donated by themselves.

Not content with that, the company has restored a 1930 Ahrens Fox, 1,250-gallons per minute pumper and has acquired a truck with a flat bed trailer to haul their antiques.

The five-year old company of Hobe Sound, Florida began painstakingly restoring a 1926 American La France pumper, a job requiring the stripping of the machine down to the chassis. In Haywood, California the firemen and their families organized to restore a 1923 Seagrave pumper. They raised money by displaying it and a borrowed antique pumper in a shopping mall. The response was tremendous. The old engine is now used in musters and competitions throughout the state.

In fact, there is a Society for the Preservation and Appreciation of Antique Motor Fire Apparatus in America. Anyone who has watched firemen's parades or antique car rallies will have seen numerous beautifully restored fire apparatus owned by volunteer fire companies.

All this is an indication that belonging to a volunteer fire company is not a sometime thing but an all-consuming passion, as many a fireman's wife will testify. Fire fighters have an almost British reverence for tradition, and a sense of belonging to an ancient and honorable calling.

Firemen cherish anecdotes about engines long since

traded in. It is not recorded that they now kiss their engines at the end of a fire as did their nineteenth-century predecessors, but they will sometimes club together to raise money to buy a cherished engine about to be sold or traded. In lists of apparatus there is often a notation about an older piece, "Held in reserve." That probably means that the company could not bear to part with it.

In an era of permissive sex and frequent divorce a lover can have little of Shakespeare's feeling that "Love alters not with his brief hours and weeks, But bears it out even to the edge of doom," or of Keats: "Bright star, would I were as steadfast as thou art," but the volunteer fireman shining up the pumper for a Fourth-of-July parade is experiencing the same love affair with the engine that his ancestor knew 150 years ago when shining up the brilliant decorations on a hand-pumped machine—also for a Fourth-of-July parade.

Today many common words have lost their former connotations or even their meaning. A student might easily misinterpret Wordsworth's "A poet could not but be gay." And there was the girl who insisted that by "Leaves of Grass" Whitman meant marijuana. But when the young fire fighter is told to hook up the steamer connection, he knows exactly what is meant, though he or she may never have seen a steam fire engine. (It is the big outlet on a hydrant.) Wherever the fire fighter lives in the United States he or she knows what it means when a fire is reported as being "a worker"—a long-established term for a sizeable blaze.

Arriving at a hydrant the man with the hose is told, as was his ancestor, "Wrap the son of a bitch around the goddamn plug." * (A woman is likely to be addressed in milder terms: firemen are old-fashioned in this respect.) The reason for the advice is that an unsecured hose will be dragged up the street. The rookies have drilled into them the ancient ad-

---

* A driver after giving such an order once asked "Who was that guy on the hose?" The answer was, "He's the new Presbyterian minister."

vice: "Never enter a burning building without a charged line."

The city boy or girl today who has never seen a sleigh outside of an antique shop can get little of the sensuous feeling of Frost's "Stopping by Woods on a Snowy Evening," or share Thomas Wolfe's tremendous sense of loneliness brought on by the early-morning clop-clop of the milkman's horse. The college student today has no visceral or neural memories of the sound of horseshoes on the pavement; they are now merely verbal symbols. But when the siren blows, the 20-year-old fireman undoubtedly experiences in his innards much the same sensations his great grandfather did when the bell in the firehouse tower began to clang. The same automatic mechanism pumps adrenalin into his arteries.

And on a warm spring evening a fire drill is likely to wind up with a hilarious water fight like those of a century or more ago.

The so-called generation gap is not very marked in a volunteer fire company. A 20-year-old may grab the hose line ahead of an elder, but he respects the veteran's know-how. By contrast, the jazz-age college youth had little respect for the Bulldog-on-the-Bank generation which in turn deplored the enthusiasm for Mencken's *American Mercury*. The parents who experienced the enforced separations during World War War II often have little sympathy for their young who casually live together. The youthful enthusiasm for unisex blue jeans and male long hair produces considerable family bickering.

There is often little sense of shared experience between elders who stuck it out through four years of college and perhaps graduate or professional study and their children who drop out to enter a commune or ashram, or between the father who has achieved business success by hard work and

his son or daughter who sneers at the materialistic rat race.
But the sons and recently the daughters of volunteer firemen
are likely to join the same company.

It is the sense of shared experience which perhaps more
than anything else holds a volunteer fire company together.
As MacLeish wrote in *Speech to Those Who Say Comrad:*

> *Who are the born brothers in truth? The puddlers*
> *Scorched by the same flame in the same foundries:*
> *Those who have spit on the same boards*
> *   with the blood in it. . . .*
>
> *Veterans out of the same ships—factories—. . . .*
>
> *Brotherhood! No word can make you brothers!*
> *Brotherhood only the brave earn by danger or*
> *Harm or bearing hurt and no other.*

Wallace Stevens was mainly right about children picking
up our bones and having no idea how we felt. But the 18-
year-old rookie fireman getting out of bed at 2 A.M. on a
winter night feels exactly as his great grandfather did and
expresses his emotion in the same language.

REFERENCES

CHAPTER I

BLOOMINGTON, MINN.: *Fire Engineering,* Feb. 1973, p. 36

BRITISH FIRE BRIGADES: Blackstone, G.V., *A History of the British Fire Service,* London, 1957, pp. 68, 98-99

de Tocqueville: Ch. 10 (Mentor, pp. 95-96)

FRENCH FIRE DEPARTMENTS: Bowen, John E., in *Fire Engineering,* Oct. 1975, p. 36

NUMBER IN VARIOUS STATES: Univ. of Minn. *Fire Department Fact Sheet,* Mar. 1976, New York Dept. of State, *IAFC Newsletter*

READING, PA.: *Fire Engineering,* Aug. 1969, p. 50

Strong:*Diary,* ed. by Nevins, Allan, & Thomas, Milton, N.Y. 1952, I, 67 & III, 6

24,000 DEPARTMENTS: *IAFC Newsletter*

CHAPTER II

BOSTON CONFLAGRATIONS: Heywood, Charles F., *General Alarm, A Dramatic Account of Fires and Firefighting in America,* N.Y. 1967, pp. 10-11

BOSTON FIRE CLUB: Bridenbaugh, Carl: *Cities in the Wilderness,* Putnam, N.Y. 1964, p. 370

BOSTON REGULATIONS: Heywood, p. 9

BOSTON: Holzman, Robert E., *The Romance of Firefighting,* N.Y. 1956, p. 14

Bridenbaugh, pp. 209-10 & 211-12;

CHARLESTON FIRES: Bridenbaugh, p. 371

CHARLESTON, S.C.: Ibid., p. 212

CHICAGO LAW: Shean, James W. and Upham, George P., *The Great Conflagration: Chicago its Past,* etc., Cincinnati, 1871

CONFLAGRATIONS IN BRITAIN: Blackstone, p. 83

DILIGENT AND HIBERNIA RULES: ms. Articles and Minutes in the Pennsylvania Historical Society

ENGINES IN 1742: Bridenbaugh, p. 371

ENGINE OUT OF REPAIR: Scharf, J. Thomas and Wescott, Thompson, *History of Philadelphia,* Philadelphia, 1884, I, 192

FRANKLIN ON PHILADELPHIA'S GOOD RECORD, *Autobiography* (Rinehart edition, p. 106)

FRANKLIN'S PROPOSAL: *Papers of Benjamin Franklin,* Yale University Press, New Haven, 1960, II, 13-14

HAND-IN-HAND CO.: Cassedy, Albert, *The Firemen's Record . . . of the History of Philadelphia. . . .* n.d. I, 146

Heywood: p. 11

James, Henry: *Notes of a Son and Brother,* N.Y. 1914, pp. 297-98

MEMBERSHIP IN FIRST COMPANY: *Our Firemen, A History of the New York Fire Departments,* N.Y. 1887, p. 28

MEN OF PROPERTY: *Autobiography,* p. 107

NEW YORK & PHILADELPHIA REGULATIONS: Bridenbaugh, pp. 207, 208, 209

NEW YORK FIRES: Wilson, James Grant, *Memorial History of the City of New York,* N.Y. 1892, pp. 526-27

NEW YORK SLOW IN FIRE PROTECTION: Bridenbaugh, pp. 370-71

PETER STUYVESANT: Sheldon, George W., *The Story of the Volunteer Fire Dept. of the City of New York,* 1882, p.1

PHILADELPHIA REQUIREMENTS: Bridenbaugh, p. 212; Scharf and Wescott, I, 192; Cassedy, p. 5

RAINBOW FIRE CO.: Burns, Robert, "Volunteers Protect 100,000 Residents of Reading, Pa.," *Fire Engineering,* Aug. 1969, p. 50

RANDOM SAMPLING: Response to questionnaire

Smedley, Elizabeth: *Local Fire Administration in Pennsylvania*, Harrisburg, 1960, p. 8

WASHINGTON AS FIREMAN: Holzman, p. 13 & pamphlet, *The Friendship Veterans Fire Engine Company*, Alexandria, Va. n.d.

WOMEN AND CHILDREN IN BUCKET LINE: McCosker, M.J., *The Historical Collection of the Insurance Company of North America*, Phila. 1967, p. 59

CHAPTER III

AUXILIARY GROUPS: Asbury, Herbert, *Ye Old Fire Laddies*, N.Y. 1930, p. 89

BALTIMORE: Murray, William A., *The Unheralded Heroes of Baltimore's Big Blazes*, Baltimore, 1969, p. 3

BOSTON ALLEYS: Forbes, Esther, *Paul Revere and the World He Lived In*, Boston, 1942, p. 163

Cassedy: p. 433

DELMONICO'S: Sheldon, pp. 128-30

ENGINE CO. NO. 42: Ibid., pp. 76-77 & 100-01

EXCUSES FOR ABSENCE: Ibid., p. 132

FATALITIES ON WAY TO FIRES: Heywood, p. 14

Hall, Captain Basil, *Travels in North America in the Years 1827 and 1828*, London, 1829, I, 81

Gulick, James and Mills, Zophar: Sheldon, p. 22

HALL'S ACCOUNT: Ibid., I, 19-22

HAZARDS AT THE BRAKES: Dunshee, Kenneth, *Enjine! Enjine!*, N.Y. 1939, p. 13

HOSE FROM HAMBURG: Sheldon, pp. 15-16

HOSE PIPE OF NO. 38: Ibid., p. 109

JENKINS, W. L.: Sheldon, p. 104

Kemble, Frances, *Records of a Girlhood*, N.Y. 1879, p. 537

Kemble: Ibid., p. 537

KLINKOWSTROM: *Baron Klinkowstrom's America, 1818-1820*, Trans. & ed. by Scott, Franklin, Evanston, Ill. 1952, pp. 76-77

LADY MOTLEY: Quoted, Holzman, p. 24

LADY WASHINGTON CO.: Dunshee, Kenneth, *As You Pass By*, N.Y. 1952, p. 34

LIBERTY CO.: *Reading's Volunteer Fire Department*, Federal Writers, 1938, p. 94

LOWER MERION: Fire Dept., *Annual Report*, 1975

MEMBERS OF HIBERNIA CO.: Sheldon, pp. 100-01, and Hibernia Minutes, Pa. Hist. Soc.

MILLS, ZOPHAR: Sheldon, p. 21

NEW YORK FIRE OF 1845: Asbury, p. 139

NEW YORK FIRE OF 1835: Morris, pp. 120-25 and *The Diary of Philip Hone, 1828-1851*, ed. by Allan Nevins, N.Y. 1936, p. 191

OTHER COMPANIES: Letters from chiefs and Clyde W. Centers, Oregon State Fire Marshal, *Annual Report for 1974*

PHILADELPHIA HOSE CO.: Cassedy, p. 27

PHILADELPHIA TYPE ENGINE: Sheldon, p. 74

READING CALLS: Burns, p. 50

RESCUE OF CANARY: Haight, James S. & Nalin, Richard L., *Brewerytown Blazes, A Century of Milwaukee Fire Fighting*, Milwaukee, 1971, p. 4

WATER IN TUBS: Heywood, p. 17

Yell, Charles, *Travels* I, 51

CHAPTER IV

BROOKLYN CONTEST: *Our Firemen, The Official History of the Brooklyn Fire Department*, Brooklyn, 1892, p. 40

BUNK ROOMS: Sheldon, pp. 21 & 149-50

CLARK, FRANCES: Asbury, p. 169

CURFEW: Morris, p. 145

DECORATION OF ENGINES: Asbury, pp. 144-45; Holzman, pp. 51-52; Sheldon, p. 67

FIGHT IN WILLARD'S HOTEL: Nicholson, Philip W., *History of the Volunteer Fire Department of the District of Columbia*, 1936, p. 22

"FIREMAN'S BRIDE,": Holzman, p. 44

GREENWOOD CEMETERY: Sheldon, pp. 125-26

Holzman: p. 35

Hone, Philip: *Diary*, p. 914

LAMENT FOR OLD DAYS: Sheldon, p. 8

L. CHAPMAN RESIGNS: Ibid., p. 5

MILLS', ZOPHAR, WEDDING NIGHT: Sheldon, p. 21

MULRINE, BETH: *Philadelphia Evening Bulletin*, Apr. 13, 1976, III, 1

NEW ORLEANS MEN IN CINCINNATI: O'Connor, Thomas, *History of the Fire Department of New Orleans*, New Orleans, 1895, p. 409

NEW ORLEANS PARADE: O'Connor, pp. 77-78

NEW ORLEANS RELIEF ASSOCIATION: O'Connor, p. 72

NEW ORLEANS TOMBS: Huber, Leonard V, *New Orleans, A Pictorial History*, New Orleans, 1971, p. 277

NEW YORK FIREMEN'S ANECDOTES: Sheldon, pp. 12, 26, 41, 99-100

NEW YORK FUND FOR THE DISABLED: Ibid., p. 112

NEW YORK MINUTE BOOK, 1813: Ibid., p. 139

NEW YORK NO. 13 HOUSING: Ibid., p. 118

n. PRINCE OF WALES: Blackstone, p. 179

n. TOWNSEND, MASS.: Lowe, Elsie L. & Robinson, David E., *Firehouse History, 1775-1975*, Lunenburg, Mass. 1975, p. 41

PARADE OF 1824: Sheldon, p. 35

PHILADELPHIA ASSOCIATION: Smedley, p. 197

PENNSYLVANIA INSURANCE COMPANIES: Smedley, pp. 197-98

PIERCE PROCESSION: C.O. Glory [pseud.] *100 Years of Glory, 1871-1971, History of District of Columbia Fire Department*, Washington, D.C. 1971, p. 10

PRIDE IN ENGINES: Sheldon, p. 103

RAINBOW RANKS: Fed. Writers, p. 2

READING COMPANY LIBRARY: Ibid., pp. 29, 98, 113

READING FIREMEN: Ibid., p. 47 & Burns, p. 81

READING TRAVELS: Ibid., p. 17

READING VISITS, 1923 & 1925: Ibid., p. 2

ROME FAMILY: Sheldon, p. 9

SHAME OF WASHING: Holzman, pp. 37 & 63

SONG: Holzman, p. 40

Strong, *Diary*, I, 68

TEMPERANCE CRUSADE: Morris: pp. 143-44 & McKelvey, Blake, *Rochester, the Water-Power City, 1812-1854*, Cambridge, Mass. 1945, p. 250

TOWNSEND, MASS.: Lowe & Robinson, p. 43

WASHINGTON, D.C.: May Day C.O. Glory pseud., p. 10

CHAPTER V

BALTIMORE FIREMEN: Murray, pp. 3, 4, & 6

. BARNICOAT, BIG BILL: Morris, p. 145

BOSTON, 1767: Heywood, pp. 15-16

CHICAGO FIRE COMPANIES: Morris, p. 118 & Pierce, Bessie Louise, *A History of Chicago*, N.Y. 1940, I, 220 & II, 312

CINCINNATI'S STEAMER: Morris, pp. 167-68

DATES OF PAID COMPANIES: Holzman, p. 103

DISORDER IN LATE 1840's: Cassedy, p. 85 & Neilly, Andrew H., *The Violent Volunteers: A History of the Volunteer Fire Department of Philadelphia, 1736-1871*, Doctoral dissertation, Univ. of Pa. 1959, p. 53

FRANKLIN AND MOYAMENSING HOSE COMPANIES: Cassedy, p. 85

MAJOR FRITZ'S SPEECH: *Historical Sketches of the Formation and Founders of the Philadelphia Hose Co.*, Philadelphia, 1854, pp. 54-55

n. BRITISH INSURANCE BRIGADES FIGHT: Blackstone, p. 71

PHILADELPHIA BALLADS: Wescott, Thompson, *History of the Philadelphia Fire Department between the Years 1701 and 1852* (made up of newspaper clippings.)

PLAY ABOUT BRODERICK: Morris, p. 149

POLITICAL POWER: Cassedy, p. 119

ROWDIES NOT ACTIVE IN NEW ORLEANS: O'Connor, p. 154

SAN FRANCISCO EXPERIENCE: Lotchin, Richard L., *San Francisco*, N.Y. 1974, p. 179

STOKLEY, WILLIAM, SPEECH: Cassedy, p. 113

TAMMANY AND GULICK: Morris, pp. 143-44

"THE SILVER HOOK AND LADDER" AND "NUMBER SIX . . . ," Holzman, p. 46

VOLUNTEERS UNTIL 1873: Handlin, Oscar, *Boston's Immigrants*, Harvard University Press, Cambridge, Mass. 1941, pp. 21 & 193

CHAPTER VI

ARSON IN BOSTON: Winsor, Justin, *Memorial History of Boston*, Boston, 1882, p. 231

ATTACK BY HEIDELBERG BRIGADE: Fed. Writers, p. 6

ATTACK ON BLACKS, 1824: Scharf and Wescott, I, 368

ATTEMPTS TO BURN PHILADELPHIA: Cassedy, p. 27

ATTEMPTS TO BURN SAN FRANCISCO: Lotchin, p. 177

BALTIMORE INCENDIARY: Murray, p. 3

BURNING OF PENNSYLVANIA HALL: Scharf and Wescott, I, 643

BURNING OF URSULINE CONVENT: Handlin, pp. 196-97

CAPT. HIESTER RAISES READING COMPANY: Fed. Writers, p. 13

DECKER, CHIEF: Morris, p. 194

DRAFT RIOTS IN NEW YORK: Sheldon, p. 82

1834 ATTACK ON BLACKS: Scharf and Wescott, I, 642

ELECTION NIGHT RIOT: Cassedy, p. 67

FIGHT FIRE NEXT TO WILLARD'S: Sheldon, p. 320; Holzman, p. 90

GOOD WILL CO. CLEARS RIOTERS: Scharf and Wescott, I, 652

HIBERNIA COMPANY IN WASHINGTON: Nicholson, p. 27

HIBERNIA HOSE CO. INCIDENT: Scharf and Wescott, I, 665-68; Morris, p. 140; Cassedy, p. 73

JUNIOR FIRE CO. FORMED: Fed. Writers, p. 27

MICHIGAN: Fire fighters Training Council, *Newsletters,* Aug. Sept. 1975

MILWAUKEE FIREMEN PREVENT RIOT: Nailen and Haight, p. 3

NEW ORLEANS FIREMEN AND BUTLER: O'Connor, p. 170

NEW YORK FIREMEN, 1776: Sheldon, p. 308

NEW YORK ZOUAVES ENLIST: Ibid., p. 308

PHILADELPHIA FIRE OF 1865: Cassedy, p. 107

RIOTS IN 1840, 1842, 1848: Ibid., pp. 67, 73, 87

SANITARY FAIR: Sheldon, p. 320

35 MAJOR RIOTS: *Philadelphia Evening Bulletin,* July 4, 1976

TOUGHS POUNCE ON FIREMEN: Morris, p. 138

ZOUAVE CASUALTIES: Ibid., p. 185

CHAPTER VII

AMOSKEAG STEAMERS: Lesure, Abner P., in *Official Journal of the Seventh Annual Tournament, New England States Veteran Firemen's League,* Sept. 1897, pp. 52-53, & Lowe and Robinson, p. 73

BALTIMORE HOSE CO. BUYS STEAMER: Murray, p. 4

CHICAGO WATER PIPES: Pierce, I, 220

FIRST AMERICAN STEAMER: King, William T., *History of the American Steam Fire Engine,* Boston, 1896, pp. 5-14

FRANKLIN PAYS: Gillingham, Harold E., "Philadelphia's First Fire Defense," *Pa. Mag. of Hist. & Bio.,* LVI, No. 1, 1932, p. 366

HAYES AERIAL: Morris, p. 160

HIBERNIA CO. MS. MINUTES, Pa. Hist. Soc.

HOSE CARRIAGES: Cassedy, pp. 37 & 55

KANSAS CITY TEAM IN PARIS: Holzman, p. 123

LYON, PATRICK: Cassedy, p. 53

NEW ORLEANS EXPERIENCE: Huber, pp. 274-75 & O'Connor, p. 117

NICHOLS', ANTHONY, ENGINE: Gillingham, p. 362

PHILADELPHIA ENGINE BUILDERS: Oberholtzer, I, 406

PHILADELPHIA GETS LATTA ENGINE: Cassedy, p. 101

PROPOSAL FOR STANDARD HOSE: Ibid., p. 27

READING FIREMEN IN PARIS: Fed. Writers, p. 64

RIVETED HOSE: Cassedy, pp. 45 & 47

ROCHESTER, N.Y.: McKelvey, p. 82
SELF-PROPELLED STEAMER IN BOSTON: Heywood, p. 40
STEAMERS IN NEW YORK: Sheldon, pp. 5 & 86
TOAST TO STEAM FIRE ENGINE: Ibid., p. 88
TWEED: Callow, Alexander, *The Tweed Ring*, Oxford University Press, 1966, p. 14 & Morris, p. 142

CHAPTER VIII
CHESTER COUNTY PLAN: *Fire Engineering*, Feb. 1973, pp. 38-39
CONNECTICUT FIRE SERVICE: *Fire Command*, Oct. 1975, pp. 21-22.
EAGLESWOOD MEN DIE: *Philadelphia Bulletin*
EARLY MOTOR APPARATUS: McCall, Walter P., *American Fire Engines since 1900*, Ellwyn, Ill. p. 8 & Lowe and Robinson, pp. 82-86
EQUIPMENT OF VARIOUS COMPANIES: letters from chiefs
FIRE ALARM INVENTED: Morris, p. 81
GARDEN STATE RACE TRACK: Meyer, Peter, "The Last Race at Garden State," *Firehouse*, July 1977, pp. 73-74
GOSHEN BUDGET: Barraclough, Robert J., "Goshen, Pennsylvania, a Volunteer Fire Company in Action," *Firehouse*, April 1978, pp. 35-36
IOWA: *Fire Service Extension*, Iowa State Univ. 1975-76
LETTER ABOUT AMOSKEAG STEAMER: Lowe & Robinson, p. 71
MICHIGAN: *Annual Report of Fire Fighters' Training School*, 1974
NEBRASKA'S FIRE SCHOOL: *History of the Nebraska State Volunteer Firemen's Association*, 1957, pp. 35-37
NEW HAMPSHIRE: Letter from N.H. Chief of Fire Training
1977 FOREST AND BUSH FIRES: *Los Angeles Times*, Aug. 18 & 27, 1977; *Philadelphia Bulletin*, Aug. 1, 1977
NORTHEAST PA. FIRE COMPANIES: Zini, Frank, "Fund Raising," *Firehouse*, Apr. 1978, pp. 67-68
OREGON PLAN: Centers, Clyde W., State Fire Marshal, *Annual Report for the Calendar Year 1974*, pp. 381-85
PENNSYLVANIA COMPANIES CROSS STATE LINES: Smedley, p. 13
PENNSYLVANIA STATE FIRE SCHOOL: Ibid., p. 229
READING'S SCHOOL: Fed. Writers, p. 126
RR. CHAIN OF COMMAND, *The New York Times*, Mar. 26, 1978
Smedley findings: pp. 19-20
TERRE HAUTE FIRE: Daniel, James, "How Good are our Volunteer Firemen," *Reader's Digest*, June 1968

VIRGINIA: *Fire Service Training*, Va., Dept. of Edu., 1975
WAVERLY FIRE: *The Tennessean*, Feb. 25, 1978; *The Philadelphia Bulletin*, Feb. 26 & 27, 1978; *The New York Times*, Feb. 26, 1978

CHAPTER IX
AFRICAN FIRE ASSOCIATION DEFEATED: Cassedy, pp. 51-53
BOYER, CAPT. LARRY: *N.Y. Times*, Mar. 23, 1975., N.J. Sect. P. 26
BRODERICK, DAVID C.: Morris, pp. 148-49
CHIEFS' LIST OF QUALIFICATIONS: Smedley, p. 150
CLOTHING PROBLEM: *Fire Engineering*, Aug. 1976, p. 19
DEL MAR CO.: *Fire Command*, Feb. 1976, p. 19
EAST LANSDOWNE ARSON: *Philadelphia Bulletin*, May 11, 1976
HAND-IN-HAND CO.: Wescott, Thompson, n.p.
INCREASE IN PA. AMBULANCE SERVICES: Smedley, p. 261
Johnson, Mike, on women firefighters: *Fire Command*, Feb. 1976, p. 17
MICANOPY, FLA.: Ibid., p. 17
MINNESOTA AMBULANCE SERVICE: *Fire Letter*, Univ. of Minn. 1976
MOLLY OF ENGINE CO. NO. 11: Sheldon, p. 47
MULRINE, BETH: *Phila. Eve. Bulletin*, Apr. 18, 1976
NEPTUNE TOWNSHIP: *N.Y. Times*, Aug. 19, 1972, N.J. Sect., p. 96
OREGON AMBULANCE SERVICE: Centers, Clyde W., p. 19
PA. JUSTICE DEPT.: *Phila. Eve. Bulletin*, Mar. 15, 1976
POINT OF NO RETURN, p. 245
PRINCETON JUNCTION: *N.Y. Times*, Mar. 23, 1975
Rambo, Carol: *Fire Engineering*, Aug. 1976, pp. 85-86
READING FIREMEN DEFEAT PAID DEPT., Fed. Writers, pp. 110, 124
TWEED'S COMPANY: Callow, p. 13
WOMEN FIRE CHIEFS: *N.Y. Times*, Mar, 23, 1975

CHAPTER X
AUTOCAR FIRE: *Phila. Eve. Bulletin*, Aug. 31, 1956 & personal experience
BRISTOL FIREBOAT: Zinn, Frank, "American Fireboats," *Firehouse*, May 1978, pp. 27-28
BUFFALO BLIZZARD: Ditzel, Paul, "Buffalo Snow Alert: Holocaust at Whitney Place," *Firehouse*, Dec. 1977, pp. 24-27 & 66
CORINTHOS FIRE: Collins, Harry, Jr., Sides, Robert, and Weldon, Wayne Curtis, *The Corinthos Disaster*, Marcus Hook, 1975

Daniel, James: *Reader's Digest,* June 1968

DONORA, PA.: Roueché, Berton, "The Fog," *New Yorker,* 1950

EATON RAPIDS FIRE: *Fire Engineering,* Aug. 1969, pp. 34-35

FACTORS IN INSURANCE RATES: Smedley, p. 41

FINANCIAL SAVINGS: Kimball, Warner Y., "Fire Departments—Volunteer and Paid," *Fireman,* Jan. 1968, p. 23

INSURANCE RATES: Letter from Insurance Information Institute, July 30, 1976

JAMESTOWN, N.D.: Letter from Chief Johnson

"JAWS OF LIFE," *Fire Command,* Oct. 1975, p. 6

LOWER MERION: 1975 *Annual Report, Fire Department,* Lower Merion Township

MONTGOMERY COUNTY, MD: Fire Command, July 1970, p. 29

NEW JERSEY OIL COMPANY FIRES: Sandell, John G., *Phila. Eve. Bulletin,* Aug. 24, 1974

1957 UNDERWRITERS' RATING: Smedley, pp. 41-42

PERCENTAGE OF MEN AVAILABLE: Kimball, p. 23

PRAIRIE FIRE: Randolph, Ruth, "Fire on the Prairie," *Firehouse,* Jan. 1977, pp. 43-45

WEST CONSHOHOCKEN FIRE: *Phila. Eve. Bulletin,* Jan. 28, 1971 & personal experience

WINSTED, CONN: John Hersey, "High Water," *The Philadelphia Bulletin* (International News Service Feature)

CHAPTER XI

EARLIER MUSTERS: Lesure, Alfred P., *History of the Fire Department of Springfield*

FRIENDSHIP CO. OF ALEXANDRIA, VA.: Booklet, *The Friendship Veteran Fire Engine Co.,* n.d. (about 1971)

HAYWOOD, CA.: Letter from chief

HOBE SOUND, FLA.: personal visit

RECENT NEW ENGLAND MUSTERS: Lowe & Robinson, pp. 14-17

REPORT OF NATIONAL COMMISSION OF FIRE PROTECTION: *Fire Command,* Oct. 1975, pp. 21-22

TOWNSEND RESTORES STEAMER: Lowe & Robinson, pp. 22-23

WAYNE, N.J. FIRE CO.: *Fire Engineering,* Sept. 1975, p. 54

INDEX

205